Rooted *in* Excellence

Rooted *in* Excellence

An Institutional History of The Arboretum at Penn State

Kim C. Steiner

THE PENNSYLVANIA STATE UNIVERSITY PRESS
UNIVERSITY PARK, PENNSYLVANIA

KEYSTONE

A KEYSTONE BOOK

Keystone Books are intended to serve the citizens of Pennsylvania. They are accessible, well-researched explorations into the history, culture, society, and environment of the Keystone State as part of the Middle Atlantic region.

The author and the publisher would like to thank The Arboretum at Penn State for its generous support in the publication of this book.

Library of Congress Cataloging-in-Publication Data

Names: Steiner, Kim C., author

Title: Rooted in excellence : an institutional history of the Arboretum at Penn State / Kim C. Steiner.

Description: University Park, Pennsylvania : The Pennsylvania State University Press, [2026] | "A Keystone book." | Includes bibliographical references and index.

Summary: "A narrative history of The Arboretum at Penn State on the University Park campus. Describes the obstacles and successes in the arboretum's planning, designing, and funding"—Provided by publisher.

Identifiers: LCCN 2026007358 | ISBN 9780271102030 hardback

Subjects: LCSH: Arboretum at Penn State—History | Arboretums—Pennsylvania—University Park—History

Classification: LCC QK480.U52 A738 2026

LC record available at https://lccn.loc.gov/2026007358

Printed in China
Published by The Pennsylvania State University Press,
University Park, PA 16802-1003

10 9 8 7 6 5 4 3 2 1

The Pennsylvania State University Press is a member of the Association of University Presses.

It is the policy of The Pennsylvania State University Press to use acid-free paper. Publications on uncoated stock satisfy the minimum requirements of American National Standard for Information Sciences—Permanence of Paper for Printed Library Material, ANSI Z39.48-1992.

GPSR Authorized Representative: Logos Europe, 9 rue Nicolas Poussin, 17000 La Rochelle, France, contact@logoseurope.eu.

Frontispiece: View over the Marsh Meadow toward the wall enclosing the Rose and Fragrance Garden (detail). Photo by Jeff Remington.

Contents

Illustrations

Preface

For most of the twentieth century, an arboretum for The Pennsylvania State University's main campus was the object of repeated initiatives and even board resolutions, all of which came to naught. For those eager to see it developed, the arboretum existed only as a dream until the century had ended and a new millennium begun. Then in 2009, a preliminary build-out of the H. O. Smith Botanic Gardens was completed at a cost of $7.3 million. As designed, the gardens were the "front door" from campus to a 370-acre arboretum that officially came into being with the event of their creation. Almost immediately, the arboretum became one of the most prominent and beloved features of the University Park campus.

The journey to success began in 1995 with a "white paper" proposal from a task force appointed by one of the University's vice presidents. This was followed by a more detailed, consultant-led plan that received the approval of the University administration in 1999. Importantly, however, action on the plan was made contingent upon success in securing an "eight-figure" gift, at least $10 million, a donation of rare size that explicitly would *not* be a fundraising priority for the administration. That such a hurdle could be overcome was not impossible. But it presented a huge challenge, and we had little reason for optimism.

Ultimately, The Arboretum at Penn State was the achievement of a small group of faculty and professional staff who persevered with a vision and, at critical junctures, benefited from the encouragement and assistance of sympathetic members of the University administration. After eight years of efforts to promote the arboretum and immediately following the quiet withdrawal of further support from the administration, success came unexpectedly with a $10 million gift from Charles H. "Skip" Smith, an alumnus whose previous donations to the University had been few and small. Skip's generosity inspired thousands of other supporters, and with donations now totaling over $60 million, the arboretum has become one of the largest privately funded projects in Penn State's history.

I have often been asked to speak to groups about the story of the arboretum, and others have suggested that the history be written down. These requests and suggestions tend to come from older fans of the arboretum who are vaguely aware that its genesis was unusual among campus building projects. One request came from Rod Kirsch, the University's senior vice president for development and alumni relations spanning the years of the arboretum's creation. To Rod, it is a remarkable story that demonstrates the power of philanthropy.

Following my retirement in 2021 as founding director of the arboretum and professor of forest biology, university archivist Ben Goldman requested that I donate my arboretum files to the University Archives. This was a very reasonable request, but it presented me with a great task. I had a habit of accumulating paperwork beyond its useful life, and I never employed a staff assistant to keep my files in order. I answered my own telephone and kept my own records. Some important documents were in the office of my longest-serving staff member, whose files were in even greater disarray. Rather than simply dump this material on the library's staff, I felt obliged to sort and sometimes cull the thousands of paper and electronic files, personal notebooks, work products, and photographs.

When finished, I was so well reminded of events and people that it seemed the best time to write the story of the arboretum, and this book is the result. As I shall explain, the arboretum underwent enormous changes in personnel and administration beginning just before my retirement and extending through the year or so afterward. These changes were so pervasive as to open a new chapter in the growth of the arboretum, so the story has a natural trajectory that coincides with the tenure of my involvement from 1994 to 2021 (or even from 1975 to 2021, if we count a previous arboretum initiative in which I played a minor role). Unsurprisingly, then, my narrative is something of a personal memoir. If, instead, I could have somehow scripted the story into a performance, it would resemble a pageant with a large and ever-changing cast of characters. The arboretum is the work of many, and I have endeavored in the narrative to credit all who were involved along the way. If it is inaccurate or incomplete in any way, then it is from oversight and not intention.

Acknowledgments

I wish to extend my gratitude to those who read and commented on earlier drafts, and especially to Kendra Boileau, Rod Kirsch, Judy Larkin, Eliza Pennypacker, Kate Reeder, Andy Stephenson, Gordon Turow, and Dave Zehngut. They reminded me of things I had forgotten, corrected errors, pointed out awkward or tedious prose, suggested improvements in the flow of topics, gently reproached me for an ungenerous passage or two, and joined me in the pleasant recollection of forgotten events. The text was much improved by their suggestions, and I am thankful for their generous help.

These individuals and others were also enormously important to the creation and growth of the arboretum over the past thirty years, and for this I shall always be grateful. Readers will encounter the names of the many arboretum staff members whose enthusiasm for the project was so integral to its success. Similarly, I endeavored to credit many others who were not part of the arboretum staff and whose assistance went beyond their merely dutiful obligations as University employees. Those people number in the scores, and some I name repeatedly in the text. Nearly everyone, I believe, is at least acknowledged in the endnotes. Deserving of special mention are Claude dePamphilis, Derek Kalp, Judy Larkin, Eliza Pennypacker, Andy Stephenson, Gordon Turow, Scott Wing, and Dave Zehngut. These people were present from the beginning and became almost constant members of the ever-changing team of staff and faculty who helped plan and build the arboretum.

Many other University employees were generous with advice and resources at critical junctures when the arboretum was barely a shoestring operation. Among these, the staff and I remember with special fondness Jeff Dice (Office of Physical Plant); Mark Wherley (The Peter R. Gould Center for Geography Education and Outreach); Ian Nalepa (Multimedia and Design Group, Smeal College of Business); James Purdum and Joy Vincent-Killian (Penn State Hospitality Services); Karen Magnuson, Jennifer Quick, Ann Taylor-Schmidt, and Joan Scholton (Department of University Publications); and James Lukens-Gable (Penn State Multimedia and Print Center).

Several other University employees played special roles. The arboretum project would have died quickly without the support and guidance of Gary Schultz, and I have been remiss if he has not heard me thank him before. Bob Steele, my early dean, encouraged my involvement with the arboretum, offered regular advice, and arranged numerous events to find our elusive eight-figure donor. Rod Kirsch took a personal interest in the project, quietly furthered it in large and small ways, and provided wise and always principled counsel. Eva Pell's enthusiasm for our work during her early membership on the Arboretum Steering Committee and the funding she provided for several graduate student projects were more helpful than she may have realized. Finally, Rod Erickson's decision to provide seed-money support in 1999 enabled everything that followed, and I felt fortunate to have his unwavering encouragement and help as provost and, later, president of the University. I am deeply thankful to these five individuals, and everyone who values the arboretum is in their debt.

Finally, I must acknowledge the arboretum's donors, who were the only truly indispensable agents in the creation of the arboretum. Skip Smith's philanthropy stands out prominently, of course, but thousands of others have also donated, and they cannot all be acknowledged by name. Even a list of only those who gave "generously" would contain hundreds of names according to most notions of a large gift, and any threshold for mention would be arbitrary. Moreover, the size of a gift is not necessarily a measure of the generosity behind it, and I hesitate to distinguish any group. We tallied donations, not generosity, and we were as touched by a young girl's gift of her year's allowance savings as we were by gifts that were far, far larger. So I will again name Skip on behalf of all of the arboretum donors, and somehow that seems appropriate. He *was* generous in giving much and asking for little in return, not even his name on a plaque, and that seems to cover all bases.

As Rod Kirsch suggested, it *is* a testament to the power of philanthropy that the arboretum was created almost solely with private donations. Our donors made success possible, and I hope it suffices to say that we truly appreciated every gift.

Rooted *in* Excellence

A Long History of Proposals and Abandoned Efforts

Before 2009, Penn State was one of only three universities (and the only land-grant university) in the Big Ten Academic Alliance without a formal arboretum or botanic garden.[1] It is not that an arboretum was unwanted. Penn State's Board of Trustees set aside land for an arboretum in the 1910s, 1920s, and 1930s and even commissioned plans (see appendix A). However, either the will or the wherewithal apparently fell short each time. None of those initiatives succeeded or even seriously got underway.

Another initiative in the 1970s also proved futile, although it set the stage for eventual success. This effort began in 1972 with Provost Russell Larson's appointment of a Penn State Arboretum Committee. Its members were Paul Pierson, Darrell Walker, Robert Nuss, Craig Oliver, and Glenn Steyers. Pierson was the University's senior landscape architect and designer for all Penn State campuses. Walker was the head of the Department of Horticulture and chair of the committee. The other three were faculty members in Horticulture or Landscape Architecture. In their January 1973 report, the committee recommended creating an arboretum and purchasing for the purpose the privately owned Mitchell Tract, seventy-six acres that lay adjacent to central campus on the northwest side of Park Avenue (appendix B). The Mitchell Tract would connect campus with University lands farther to the north and west that were also proposed for inclusion in the arboretum.

In April 1975, a few months after arriving at Penn State as an assistant professor in the School of Forest Resources (now Ecosystem Science and Management), I received a letter from Provost Larson appointing me to the Arboretum Committee. In his letter, Larson stated that "the plan to develop an arboretum is a matter of considerable interest to the University." Serving on the committee aligned perfectly with my professional interests, and I had even interviewed the previous year for a scientist position at the Cary Arboretum of the New York Botanical Garden. I was eager to participate.

When I joined, the committee had already fulfilled its main charge from the provost, which was the report. To my knowledge, the committee had no new directions from him, and I do not know why he nominally strengthened it by adding me as a member. We met at irregular intervals over vague agendas, and meeting minutes were never distributed. Especially after Provost Larson's retirement in 1977, the committee seemed to operate under a hollow franchise motivated by the hope that something not under its control would happen to move the arboretum proposal forward. My file on the committee essentially ends with a thirteen-page promotional booklet created in 1980. This was my unfortunate introduction to academic committee assignments, and I am sure I learned more from the experience than the University gained from my service.

In November 1988, a landscape architect from Pittsburgh, Geoffrey Rausch, presented a campus seminar on "arboretum planning" at the invitation of Glenn Steyers.[2] The event took place in the Physical Plant Building, somewhat distant from central campus, and was intended for a limited audience. Committee members, planning staff in the Office of Physical Plant (OPP), and some senior administrators were invited. Among them were several people whom I will introduce shortly (Gary Schultz, James Wagner, Steve Wallner, and Jim Starling). I had a schedule conflict and do not know who actually attended. Not long afterward, in January 1989, the University purchased the Mitchell Tract. The property was undoubtedly a high-priority purchase for the University, regardless of its suitability for the arboretum as proposed sixteen years earlier, but the timing of the seminar and the purchase is interesting. I think it is possible that Steyers arranged the event after learning that the University was in negotiations to buy the property. Regardless of whether they were connected, those two events helped set the stage for what followed.

Ambitions for an arboretum came to life again in October 1994 when Jim Wagner, vice president for business and operations, appointed an Arboretum Task Force composed of faculty and staff. The task force was asked to evaluate the case for an arboretum and identify possible locations. There were ten members, and Steve Wallner (professor and head, Department of Horticulture) was appointed chair.[3]

What triggered this action by Wagner is somewhat uncertain, although I've queried several people who may have known and forgotten, including Wagner himself, Gary Schultz, and the then–dean of the College of Agricultural Sciences, Lamartine Hood. However, in December 1993, five department heads in the College of Agricultural Sciences had cosigned a letter to the administration advocating for the creation of an arboretum to address a "major unmet need." Steve Wallner was the author, I believe, and other signatories included the department heads of Entomology, Plant Pathology, and Agronomy: Jim Frazier, Elwin Stewart, and Al Turgeon.[4] I signed the letter as interim director of the School of Forest Resources. When I asked him recently, Jim Wagner could recall no champion for the arboretum within the administration but said that there was "little or no opposition" to the task force, so I believe it was probably created as a straightforward response to our petition.

In his charge letter and verbal remarks during our first meeting, in November 1994, Wagner referred to broad-based interest in an arboretum as an academic facility and "green belt" of preserved open space. (He retired shortly afterward, and his replacement, Bill McKinnon, also described the arboretum as a potential open-space buffer zone when he met with us the following January.) We learned that the University had committed to developing a new master plan for the University Park campus and that the work of the task force would contribute to that plan. In general terms, Wagner described a possible arboretum footprint drawn from the 1973 report. He emphasized that there was no guarantee that the University would act on our recommendations when we finished.

The task force met every two weeks until the end of April 1995. We divided responsibilities, and everyone contributed through outside taskwork. Carl Keener (faculty, Department of Biology) and I traveled to the University of Pennsylvania's Morris Arboretum to meet with its director, Paul Meyer, whom I knew. Meyer gave us helpful advice on justifying, programming, budgeting, and administering an arboretum within a university structure. The task force met with two Centre Regional Planning Agency members, Tom Zilla and Bob Bini, who summarized discussions about a proposed "Eastern Inner Loop" (about which I will say much more below). They believed our report could further their work of regional planning for open space and transportation needs.

In another meeting, the task force met with Joab Thomas to seek his advice as president of the University and as a botanist who had studied at Harvard's Arnold Arboretum. Naturally, his comments focused on advice about what we needed to do to make an effective case. However, I particularly remember his admonition to be ambitious, stating that the University would be better off with no arboretum than a poor one. I thought this made great sense, and it later guided our decisions as planning and construction progressed. In particular, President Thomas's words helped me resist suggestions to settle for less than we wanted, for "less" could have been as little as a park with labeled trees.

Our finished report was sent to Bill McKinnon, Gary Schultz (senior vice president for finance and business), Joab Thomas, and John Brighton (provost) on July 3, 1995, recommending the creation of an arboretum on 370 acres of University-owned farmland, starting at the Mitchell Tract along Park Avenue.[5] Two weeks later, Schultz acknowledged receiving the report in a memo to committee members. Although we did not know it at the time, he had also directed his administrative assistant, Joan Coble, to schedule a breakfast meeting with Bill McKinnon and our chair, Steve Wallner, to discuss the report. However, just two days before our report was submitted, Wallner left to take a new position at Colorado State University. Of course, with him gone, the breakfast meeting never occurred.

After several months, I invited the members of the task force to a meeting to find out whether anyone had heard more about the fate of our report. No one had new information, and after some discussion we agreed to request a meeting with Schultz. I had come to the meeting hoping that Eliza Pennypacker (faculty, Department of Landscape Architecture) would assume the leadership of our group. As an implicit offer of the role, I suggested that she convey the request to Old Main. She demurred, citing a heavy workload, and none of the others volunteered for the task, their silence probably encouraged by the knowledge that as the meeting's convener, I would feel obligated to carry through.

Had anyone else stepped forward, I believe that person (or someone else) would have become the central figure in the arboretum project. I liked my teaching assignment and had graduate students, a well-funded research program, and several important professional obligations outside the University. Just then, those responsibilities needed special attention because I was coming off my temporary assignment as director of the School of Forest Resources. I was well situated professionally, and putting effort into an arboretum quest offered no benefit to my career. Naively, I did not anticipate that by seeking a meeting with Schultz, my continued leadership would become almost inevitable by degrees, as with the proverbial boiled frog. I certainly would have declined at the outset if I had known how much time the project would consume as the cause remained uncertain, and even doubtful, for twelve more years. But "in for a penny, in for a pound," I succumbed to the fallacy of sunk costs and would not quit.

I called Joan Coble on October 24, and she quickly and graciously scheduled a meeting in December between task force members and Schultz, McKinnon, and Bill Anderson (assistant vice president for physical plant). Implementing the task force's recommendations would affect my academic college, Agricultural Sciences, so I met in November with Jim Starling, the dean, to get his response to our report. He strongly favored the project, especially if it were developed on the Mitchell Tract. However, he was concerned that relinquishing lands beyond the Mitchell Tract would adversely affect the Department of Dairy and Animal Science. Starling made the same points in a memo to Gary Schultz on November 29, with the caveat that areas beyond Big Hollow could be released for arboretum use as the College's needs changed or were met at alternative sites. Shortly after this memo, our meeting with Schultz et al. was rescheduled for January 5, 1996, and the list of participants was expanded to include Starling, Neil Porterfield (dean, College of Arts and Architecture), and Provost Brighton.

All members of the task force attended the January meeting except for Steve Wallner, Perry Morgan (who also had left the University), and George Dickie. We tag-teamed a presentation of our report's highlights. Support for an arboretum seemed unanimous among the administrators. Brighton and Anderson each stated that a formal plan and cost-benefit analysis should be developed before attempting to find funding, which all participants assumed would necessarily come from philanthropy. Schultz volunteered that OPP would cover the cost of a planning study. There was consensus that the proposed site should be set aside for arboretum use in the forthcoming campus master plan ("we have talked about this long enough") and that an arboretum planning study would provide the detail for that spatial set-aside. We could not have hoped for a better outcome.

In a memo to the provost after the meeting, Dean Starling reiterated his "enthusiastic support for creating a Penn State arboretum." He committed the College of Agricultural Sciences to vacating the Mitchell Tract (it had leased the tract long before the 1989 purchase), and he stated that lands beyond the Mitchell Tract could be released to the arboretum under the conditions laid out in his November memo to Schultz. Because the college's programs would benefit from an arboretum, he said that Agricultural Sciences would be willing to "bear a special burden" by giving up land and taking "other steps needed for a successful outcome." This was an important commitment, and the college's burden did become significant. In addition to relinquishing land, it eventually provided office space for arboretum staff, sponsorship of fundraising events, and administrative support functions for facilities, HR functions, and budget operations.

The University Commissions a Formal Plan

Gary Schultz memorialized in writing the key outcomes of the January 5, 1996, meeting and directed Bill Anderson to work with the task force to retain a consultant for the arboretum plan. We acted immediately. Following input from other task force members, especially Ken Tamminga (faculty, Department of Landscape Architecture), I sent a memo to Gary on January 17 outlining what we believed to be the appropriate scope for a consultant-led study: developing the program and a diagrammatic plan for an arboretum, estimating construction costs and operating budgets, and proposing an administrative structure. Gary accepted our recommendation.

Two weeks later, Ken, Dave Zehngut (architect and later university architect, OPP), John Joseph (senior landscape architect, OPP), and I met to develop a list of qualified firms and prepare a request for proposals. From among the seven firms that eventually responded to our RFP, we interviewed three in early June and selected Sasaki Associates of Watertown, Massachusetts, for the project. Sasaki's team was led by landscape architects Joe Hibbard and Dave Mittelstadt, whose experience included developing plans for arboreta at Harvard and the Universities of Wisconsin, North Carolina, and Illinois. As we refined the project through discussions with Hibbard and Mittelstadt, we added consultations with four arboretum directors: Greg Armstrong (University of Wisconsin–Madison Arboretum), George Briggs (The North Carolina Arboretum, affiliated with UNC), Tony Endress (University of Illinois Arboretum), and Gerry Donnelly (The Morton Arboretum).

The Sasaki master plan was the first of over a dozen arboretum planning projects for which we hired consultants over the next twenty-five years. OPP administered these contracts through a project manager, with additional oversight from one or more design professionals on its staff. OPP's design group became managed by Eliza Pennypacker (temporarily taking a leave from her faculty position in Landscape Architecture), and in 2000 it was organized as the Office of Campus Planning and Design. Dave Zehngut and John Joseph provided design oversight for this first project, and Dave also served as project manager. John retired soon after we completed the Sasaki project, Eliza returned to her faculty position, and the University hired Gordon Turow (also a landscape architect) to succeed Eliza as director of Campus Planning and Design. Dave and Gordon were key members of every arboretum design project until their retirements in the late 2010s, and Eliza participated in nearly all of them.

The arboretum was the "client" in these projects, which positioned me as the arbiter when one was needed. Experience has taught me to have high confidence in my judgments, but I tend to avoid reaching a firm conclusion until all perspectives and ideas have been aired. From the beginning, I took steps to get input for our plans from as many interested faculty and staff members as possible—and very often from non-University people, when we knew whom to ask—in the belief that success would require ideas and support from the broader Penn State community. The large size and diverse composition of these planning teams were often unusual for such University projects, but no one objected. Together, we became active advocates for good design and Penn State's interests. I think our assertiveness sometimes challenged consultants, but almost all of them responded positively to those challenges in an atmosphere of mutual respect. Dave Zehngut, John Joseph, Eliza Pennypacker, Gordon Turow, and eventually Steve Watson (a later director of Campus Planning and Design) were invaluable partners in these endeavors.

For the Sasaki contract, we scheduled on-campus workshops in February, April, and November 1997, each involving up to a dozen meetings and work sessions with various groups. Most original Arboretum Task Force members participated, and we added several more faculty members to form what I will call the Sasaki plan oversight team.[1] In addition to oversight team members, meeting participants included members of the University's central administration (including the new president, Graham Spanier); faculty and administrators from seven academic colleges; representatives from Johnson, Johnson & Roy (JJ&R), the firm working on the University Park campus master plan; members of the University's Facilities Planning and Advisory Board;[2] personnel from the Office of University Development; relevant professionals from OPP; and community leaders, including representatives from four local planning commissions, the State College Area School District, the College Heights Neighborhood Association, the Centre County Historical Society, Clearwater Conservancy, and the Centre Region Park and Recreation Authority. In the aggregate, attendance lists included over eighty names, and many people participated in all three workshops.

The administration asked us to consider the inclusion of space in the arboretum for the Department of Agronomy's turfgrass research plots. These research plots were expected to be displaced by a project (described below) to widen and extend University Drive north of Park Avenue. Sasaki suggested creating an "Overlook Heights Teaching and Research Area" in a corner of the arboretum where the turfgrass facility would fit well. The prospect of collaborating with the highly regarded turfgrass program suggested the opportunity for a robust public outreach function in the future arboretum. After meetings with Tom Watschke and George Hamilton (faculty, Department of Agronomy), a turfgrass research facility was incorporated into the plan to occupy most of the arboretum acreage next to the Overlook Heights residential area.

A gift large enough to move the turfgrass facility never came, the research plots were never displaced by construction, and the idea died a natural death within a few years. I mention all this partly to explain a significant feature of Sasaki's master plan, which is presented in appendix C, but also because abandoning the turfgrass research facility left no clear plan for future arboretum use of the agricultural fields that were (and still are) in the Overlook Heights section of the property. Because of the tract's possible value for non-arboretum (and non-agricultural) purposes, a program for this area remains the most important unresolved question for long-term arboretum planning.

As it happened, alternative uses for the arboretum tract were proposed or under consideration even as the Sasaki project was underway. All of them would have required property carve-outs and otherwise degraded the qualities that made the proposed location so suitable for an arboretum. In the face of these proposals, the members of the oversight team were naturally protective and even defensive on behalf of the future arboretum.

A plan had been advanced in 1996 for the University to build a retirement village that would have special appeal to alumni and others affiliated with the University, a project that eventually became The Village at Penn State. One location proposed for the Village was a large portion of the arboretum tract adjacent to the Overlook Heights neighborhood. I first learned of this not through University channels but from a report in the *Centre Daily Times*—and just weeks after the January meeting when we got the nod to plan an arboretum for this space. To gather more information and express my concern, I phoned George McMurtry, the chair of the Facilities Planning and Advisory Board (FPAB). He told me that the committee had already objected to using the arboretum site for this purpose, and he sent me additional information by email. Following a May 1996 meeting of the FPAB (which I attended), McMurtry reported to Bill Anderson that his committee continued to oppose the arboretum site for a life care facility. In March 1997, the new chair of FPAB sent another memo to Bill Anderson expressing the committee's concerns about "intrusions" on the arboretum site, referring to the retirement facility and to a new roadway described below.

As I said, the oversight team felt very protective of "arboretum land," and sometimes this found expression in gallows humor. At one of our meetings, in the summer of 1997, Eliza Pennypacker made the tongue-in-cheek suggestion that if a piece of the arboretum were to get carved out for a retirement village, then maybe we should fund the arboretum by selling burial plots in "an alumni cemetery" elsewhere on the property. We shared another laugh one week later when Eliza's wry suggestion was perfectly channeled by a *Doonesbury* comic strip. The strip depicted an aide briefing President King of Walden College about their fundraising efforts. The punch line was that the college was doing well with alumni subscribers to *their* retirement village, but the Walden College Cemetery was "still a little slow."[3] Later that year, we learned that the Village would be built on the north side of Interstate 99, near the arboretum but not *in* it, so we spoke no more about the Pennypacker/King business scheme.

Also in 1996, the Center for Sustainability had been directed to move from its location on the University's Circleville Farm property, and the arboretum was suggested as a potential site. Accordingly, we invited Barbara Siebert, the center's director, to attend a session during the February 1997 arboretum planning workshops. She followed up with a memo proposing that the Center for Sustainability be physically (not administratively) housed within the future arboretum, but she did not suggest a specific location. The oversight team supported the idea in principle, and I envisioned the possibility of future collaboration and a partnership.

The center then proposed to occupy twenty-five acres at the front of the Mitchell Tract, near Park Avenue. This was oddly naive, or perhaps presumptuous, because it would require us to relinquish the area closest to campus and most central to the 1973 and 1995 arboretum proposals and, of course, the emerging Sasaki plan. I told Siebert that her proposal was unworkable and later met with her and Chris Uhl to discuss alternatives. Uhl was a faculty member in Biology and an advocate for the center, and he was also a member of the oversight team. During the second meeting, I learned that they had already selected another location within the arboretum, but this, too, had already been programmed by Sasaki for a significant arboretum feature. Siebert and Uhl did not want to discuss alternatives, and their insistence suggested a degree of inflexibility that did not augur well for a cooperative, give-and-take arrangement.

I was also very concerned about the implications of the center's occupying part of the land before the arboretum even existed, and I soon learned that FPAB had already taken a position against moving the center to the arboretum. At this point, I suggested a location behind the auxiliary buildings along the upper portion of Big Hollow Road, adjacent to but not in the arboretum footprint as diagrammed in Sasaki's developing plan. OPP assigned this site to them in September 1997. For reasons unrelated to the arboretum, they were directed in 2000 to move once more. The College of Agricultural Sciences was asked to find a location for them on college lands, and the responsibility fell to Paul Wangsness, the college's senior associate dean. Wangsness consulted me, perhaps hoping there might still be a chance to put the center in the arboretum. I proposed a piece of some land near Porter Road that had been assigned to my use for research purposes many years before, but which I no longer needed. That became the location of the Center for Sustainability, and it still is.

As Sasaki and JJ&R worked on their tasks, a proposal to construct an "Eastern Inner Loop" transportation corridor emerged as a prominent topic in our meetings with University administrators and community members. This idea, which had been around for twenty-five or thirty years, was given new life in 1997 when Centre County, State College Borough, adjacent townships, and the Pennsylvania Department of Transportation formed a partnership to finance engineering and environmental studies of the route. The new road was envisioned as an extension of University Drive that would connect to North Atherton Street by way of Vairo Boulevard. As such, it would cut directly through the land intended for arboretum use. Naturally, the project sparked much discussion within the Centre Region community. Support was strong among residents whose daily commute would be made easier, while opposition focused on the negative impacts on the arboretum and nearby natural areas. Apart from finding money to build the arboretum, the Eastern Inner Loop was without question the most worrisome challenge to our project in the late 1990s and early 2000s.

The University administration did not favor the Eastern Inner Loop, but it was not opposed to the road. In comments made in my presence, Graham Spanier and Gary Schultz acknowledged that Penn State could not stand in the way of the community's solution for alleviating congestion that was mainly caused by traffic generated by the University. Their stance was to support the idea, in concept, and cooperate in developing the road if its need were demonstrated and its construction justified. In a July 1997 draft of their report, the Sasaki team strongly criticized a route that would bisect the arboretum. This criticism became a central topic of discussion when Dave Zehngut and I met with Graham and Gary in April 1998, and Sasaki's tone was considerably softened in its final report.

If there were any strong advocates for the Eastern Inner Loop among community leaders, they did not speak out in our planning workshops. Some spoke *against* the idea of traversing the arboretum tract with a new road, while others agreed that the arboretum introduced a new variable to balance against transportation demands. This topic was the subject of an article in the *Centre Daily Times* ("Inner Loop, Arboretum Plans May Conflict") on December 21, 1997. The article presented the issue as a road vs. arboretum proposition, and its last paragraph—perhaps reflecting an intended slant—was a quote from me on behalf of open space preservation. On January 7, 1998, the newspaper's editorial summarized the issue as "two worthwhile projects test[ing] town-gown relations" and did not take sides. The arboretum was strongly defended in letters to the editor on January 15 by Bruce Lindsay (a Penn State faculty member and arboretum supporter) and January 27 by Jean Aron (a well-known local hiking enthusiast).

The Eastern Inner Loop continued to be a threat to the arboretum for several years after the Sasaki plan was completed. However, political support for the Eastern Inner Loop gradually diminished as estimated costs escalated and concerns persisted that the road would degrade the environment of Big Hollow and the arboretum. On April 2, 2004, a *Centre Daily Times* editorial argued strongly against the project because the original rationale for the road was greatly weakened by the completion of the I-99 bypass around State College.

By October 2005, College Township and State College Borough had canceled their support for the Eastern Inner Loop. Jack Jenkins, a member of the College Township Council, was a particularly vocal advocate for protecting the arboretum and disapproved of the proposed University Drive / Vairo Boulevard route. He had visited me in February 2002 to "gather ammunition," as he said, and I described the impact of the route on the Sasaki plan. As chair of the council, Jenkins's opinion was likely influential. When College Township withdrew, the project was effectively dead, because most of the proposed route, especially the most controversial section, lay within the township's boundaries. The Centre Region Council of Governments voted overwhelmingly in November 2005 to terminate the project, and the Eastern Inner Loop idea has never since been seriously revived.

I return now to the arboretum master plan as its development continued in 1997. As noted above, our consultants completed a preliminary draft in July. This was revised over the next several weeks after feedback from the oversight team and Gary Schultz, who passed along compiled comments from Graham Spanier, John Brighton, Bill Anderson, and Bill McKinnon. Additional feedback was gathered during the six meetings of Sasaki's November workshop visit, and we undertook yet another revision to address those comments. At the same time, Sasaki was busy developing projections of capital costs and a proposal for administering and staffing the arboretum.

Shortly after receiving a revised draft from Sasaki in March 1998, Dave Zehngut, Ken Tamminga, and I presented the plan orally in a meeting with Graham Spanier, Gary Schultz, John Brighton, Bill Anderson, Bill McKinnon, and Charlie Brueggebors (university architect, OPP). They had read the plan in advance and requested a few moderately substantive changes and additions, the most significant being a directive from Graham to present construction costs, operational costs, and precise staffing levels in the form of a ten-year phasing plan. We were told that for action by the Board of Trustees, the arboretum plan would be subsumed within the University Park campus master plan. That plan was expected to be completed in about nine months, and this became our deadline to finish all remaining work.

The information Graham requested was outside the project's scope. However, the numbers we needed were presumably embedded within information that was already available, so rather than pursue an amendment to the contract, Dave, Ken, and I decided to undertake the ten-year plan ourselves and seek input on phasing options from Hibbard and Mittelstadt. This task proved time-consuming and frustrating because of incomplete data and obscure calculations from a third party that had been subcontracted to project a maintenance budget. We eventually got what we needed but did not finish until November. As the deadline approached, we met for weekend work sessions to complete the job.

The final report, including the landscape plan shown in appendix C, was sent to Gary Schultz, Bill Anderson, and Charlie Brueggebors in January 1999. The full report consists of fifteen pages of narrative, twelve pages of background data, and ten tables showing annual costs for initial development and subsequent phases.[4] The estimated total cost of construction, in 1999 dollars, was $31.2 million.

From Plan to Action, on a Wing and a Prayer

Although it would be another decade before the arboretum opened, 1999 stands out as an eventful and pivotal year in its development. Several important meetings lay ahead after we finished the master plan. In preparation for these, Dave Zehngut and I met in late January with Gary Schultz and Dan Sieminski (director of finance and business) to review the budget figures in the master plan and to get advice on what to expect. In the following weeks, we gave presentations to two committees advising the administration: the Facilities Planning and Advisory Board and the Faculty Senate Committee on University Planning.

At its mid-March meeting, the University Board of Trustees voted to approve the new University Park campus master plan, including, of course, a 370-acre set-aside for the arboretum. Board approval had been expected, but it was a necessary predicate for the administration to act on the Sasaki report.

The critical meetings for us took place in rapid succession soon after the board meeting: on April 5 with the President's Council (essentially the provost, vice provosts, and vice presidents) and on April 12 with the Council of Academic Deans. Dave, Ken Tamminga, and I gave coordinated presentations to each group. The meeting with the President's Council was the most important, because that was when we hoped to receive approval to proceed with the arboretum project. However, we also hoped to cultivate campus-wide support from our academic deans.

Our presentations to the President's Council generated a number of questions and some discussion. The only question I remember clearly—perhaps because it was direct and carried the whiff of a challenge—came from Tim Curley, the athletic director, who wanted to know how the arboretum would compare with those at other universities, especially those within the Big Ten. Not wanting to appear to exaggerate, I checked my swing for a bunt and said it would compare very favorably, which could have meant only that it would be "nice." That was more modest than I believed.

President Spanier gave his blessing to the plan, but he stipulated that the project could not proceed without a lead gift in the "eight figures," the minimum for a core facility according to our phasing options. He emphasized that until we had such a gift, we must not solicit smaller donations, because they could never be used as promised if sufficient start-up funds did not materialize. This stipulation was principled, but it presented us with a very high hurdle to overcome. Importantly, we did not receive an unconditional commitment to proceed, which would have implicitly carried the promise of University resources if fundraising came up short.

Gary Schultz summarized the discussion, stating that the project would not begin without a gift of at least $10 million. He also stated that we would probably need an area plan for the Mitchell Tract, where the most intensive development was planned. Graham directed Rod Kirsch (senior vice president for development and alumni relations) to "seek out the funds," which I learned later was not as straightforward a directive as it appeared.

On April 27, Joe Hibbard and I presented the arboretum master plan in two public meetings, one in the afternoon at the Palmer Museum of Art auditorium and the other in the evening at the State College Borough Building. Attendance was heavy at both. For the first time, several people approached me after the presentations with offers of donations. I had no clear idea about how to receive a donation, but the Office of University Development found an allocation code for arboretum donations that led to an old fund, probably created in the 1970s or 1980s when Provost Larson's committee was operating. The account held $348. The first new donation for the arboretum occurred on May 3, 1999, when Bruce Lindsay emailed me to say that he and his wife, Laura Simon, had just declared Penn State payroll-deduction gifts totaling $50 per month. Many other unsolicited donations followed. Bruce and Laura's charity to the arboretum continued in other ways, and in August 2015 their nine-year-old daughter, Maia Lindsay, made the local news by donating a year's worth of allowance savings to the Children's Garden. Sadly, Bruce had passed away earlier that same year.

Our plans for the arboretum generated quite a bit of publicity in the local news, beginning even before the master plan was completed. The first article appeared on the front page of the November 13, 1997, issue of *Intercom*, the University's weekly employee newsletter. Entitled "Arboretum Plans Still on Drawing Board," the article provided a lengthy description—forty-four column inches—of our vision for the arboretum. It was favorable publicity and provided encouragement to those of us on the oversight team. This was followed on December 7 by a front-page article in the *Centre Daily Times* ("Penn State Arboretum Plan Cultivates Interest"). After the plan was released in 1999, more articles appeared in the *Intercom* ("Proposed Arboretum Detailed at Public Meetings," May 6), the *Centre Daily Times* ("Arboretum Part of PSU's Grand Plan," July 10), the *Daily Collegian* ("Arboretum Plans in Works," October 26), and again in the *Intercom* ("Arboretum to Progress Under Professor's Direction," October 28). There were many more feature articles in these and other publications over the next decade or so—in *Town & Gown*, *Research*

Penn State, *The Penn Stater*, *Harrisburg Patriot-News*, and even *Kickoff*, Penn State football's game-day program. We did not pitch those articles, so it is fair to say that they both reflected and fueled public interest in the project.

Two key recommendations from the arboretum directors hired as consultants for the Sasaki plan were to immediately appoint an arboretum director and to ensure that he or she should report to the highest executive administrator within the University. The rationale for the second recommendation was that academic departments and colleges were poor homes for first-class public gardens, in part because of budget capacity but mainly because their core academic priorities tended to favor investment elsewhere. Gary Schultz seemed supportive of both recommendations when we met on April 30 to discuss the next steps for the arboretum project.

Since August 1997, I had kept the new Agricultural Sciences dean, Bob Steele, informed about progress with the arboretum plans. He was familiar with the University of Wisconsin's arboretum and its value to the university. He was enthusiastic about our project and pursued the question of arboretum leadership in discussions with Gary, John Brighton, and (after July) the new provost, Rod Erickson. It became clear through these discussions that there would be no funding to hire a full-time director.

The matter of a director was settled in a meeting on October 1, 1999, with Rod Erickson, Gary Schultz, Bob Steele, Rod Kirsch, and Bill Anderson. Rod Erickson appointed me as arboretum director and provided a stipend sufficient to support twelve weeks of my salary, a half-time staff assistant, and incidental expenses. With the second of the consultants' recommendations in mind, I asked to report to the provost in this role. Rod declined, however, citing the fact that he already had (as I recall) something like thirty-eight people reporting to him. As a compromise, it was decided that I would report to an administrative council of deans (subsequently called a steering committee) chaired by Dean Steele, with the line of administrative authority extending from the provost to the steering committee to me. A few weeks later, I paid courtesy visits to those who would be appointed to this committee: the vice president for research (Eva Pell) and the deans of Eberly College of Science (Dan Larson), Engineering (Dave Wormley), Arts and Architecture (Neil Porterfield), and Earth and Mineral Sciences (John Dutton). Bob's staff assistant scheduled the first meeting of the Arboretum Steering Committee for January 19, 2000, to discuss "arboretum implementation planning."

The Arboretum Steering Committee operated from 2000 to 2021, when a new administrative structure was created following my retirement. Committee membership evolved as individuals retired or left the University.[1] On two occasions, membership was adjusted by adding or dropping a college based on perceived interest. To create a stronger connection between the provost and the arboretum director, committee leadership transitioned in 2014 from the dean of the College of Agricultural Sciences to the vice provost for academic affairs.

The Arboretum Steering Committee did not function as an executive committee. This would have been impractical and perhaps, from the deans' perspectives, unwanted. For many years, there was no arboretum and almost no staff, and the job was largely routine even when there finally became something to "direct." When guidance was needed, especially when issues arose that might have broader implications within the University, I consulted the committee chair or the provost. That said, I kept the committee informed of progress and consulted it particularly on policy matters, and the committee was a valuable source of advice and institutional knowledge.

The committee occasionally provided welcome support through resources or the means to acquire them. Particularly helpful were assistantships offered by Eva Pell through the Graduate School to support several student projects, some of which became indispensable resources as the arboretum grew.[2] John Dutton provided the services of his staff to create our first website. The committee also leveraged supplemental funds from Penn State's Environmental Consortium (later the Institute of Energy and the Environment) to strengthen the arboretum staff. As described below, Bob Steele (who chaired the committee until his retirement in 2009) was proactive in our search for a leadership gift and generous with the resources he had available for development activities. In later years, committee members pledged sufficient financial support to hire a public programs manager for three years.

Once the arboretum became an official project, I thought it desirable to broaden faculty involvement beyond the few disciplines represented by the Sasaki plan oversight team and its predecessor, the Arboretum Task Force. In February 2000, I solicited University Park deans for the names of potential members, and a Faculty and Staff Advisory Committee was soon formed with twenty-two members representing nine academic colleges, the University Libraries, and OPP. As I had hoped, most members of the Sasaki plan oversight team were willing to transition to this new committee. We met for the first time in April.

The first major project for this committee was a master plan for the Mitchell Tract. We met periodically for this and similar purposes over many years. The committee's contributions shifted from the abstract to the concrete in 2007, when we began planning our first construction project and the committee's recommendations began to have on-the-ground consequences. This change brought heavier responsibility to the committee, and I asked who wished to continue. Some took the opportunity to leave, and I recruited others to take their places. By 2011, though, we became so heavily involved in planning new projects that our needs were better suited by ad hoc committees staffed according to interest and expertise. The advisory committee had become obsolete, but many members continued their involvement on special planning teams or standing committees.[3]

Keeping the Faith Through Uncertain Times

The 1995 Arboretum Task Force report had recommended that the arboretum should be a featured objective in the upcoming "Grand Destiny" capital campaign (1996–2003). I asked about this repeatedly in meetings with senior administrators until I was told, with finality, that the arboretum would *not* be a University priority in the campaign. The University was about to undertake what became the most rapid and costly period of new construction in its history, and the president's focus would be finding the money to pay for *those* costs.

This is the qualification I mentioned earlier regarding the president's directive to Rod Kirsch to seek out the funds. As Rod explained it to me, the development staff would respond with information about the arboretum if a donor inquired, or they might suggest the arboretum if none of the University's featured objectives appealed to a donor with whom they were in discussions. However, prospective donors known to have an eight-figure potential would not be solicited to give to the arboretum, and that excluded a large fraction of a small pool of realistic prospects, the rest of whom by definition were unknown. Our only hope was that someone with the means would come forward on his or her own initiative.

Three interested parties came to our attention in 2000 as potential donors. In May, Ed and Helen Hintz informed Rod Kirsch that they were considering a leadership gift to the arboretum. Ed was the volunteer chair of the "Grand Destiny" campaign and served as chair of the Penn State Board of Trustees from 2001 to 2004. The Hintzes have been major donors to the University in various ways and later made the largest gift for the arboretum's Children's Garden. They did not end up giving the leadership gift we hoped for in 2000, but their early interest gave me hope.

The second potential donor came to light in August. Pat Roenigk, director of planned giving and endowments, received a phone call from the financial adviser to an anonymous client. The client wanted to know whether the arboretum could be named in recognition of a gift of $8 to $10 million. We presented a couple of proposals, one in September and another in October, and felt very optimistic. But the anonymous client fell silent in mid-November, when the stock market turned against some of his investments.

The third potential donor was Peter Tombros, who, we learned, was interested in making a leadership gift to the arboretum. Peter was a College of Agricultural Sciences alumnus, and Bob Steele arranged for us to meet in November 2000. Discussions with him about donating to the arboretum continued from early 2001 to 2005. Peter and his wife, Ann, were eventually among the biggest donors in the history of the University, but we never found a formula that matched the arboretum's needs with their interests.

These early leads were very encouraging, but they held as much promise of ultimate success as a lottery ticket that is one digit shy of the winning number. Even in 2000, a year after the excitement generated by the Sasaki plan, I was beginning to understand that if a gift did not materialize soon, our prospects would fade unless the project were kept fresh in the public mind. Here I took a cue in patience from my late brother, whom I liked to accuse of never wanting to come back *too early* from our fishing trips on the river. He was always sure that fish would be biting at the next spot he had in mind. For the next seven years, until the lead gift was secured, I saw my principal task as keeping the arboretum idea sufficiently alive and appealing to catch the attention of an unknown donor while encouraging optimism in the other project advocates who were sharing the boat with me.

Between 2000 and 2007 I gave dozens of presentations about the arboretum to groups of fewer than ten to many hundreds. Other talks were presented by faculty colleagues, notably Andy Stephenson (Department of Biology) and Jim Sellmer (Department of Horticulture), who spoke to groups with which they were affiliated. Our presentations were usually made at the request of local organizations such as church groups, service clubs, neighborhood associations, retirement communities, alumni groups, academic departments, and garden clubs. Bob Steele and his staff organized several events to reach specific donors or donor groups. In March 2000 and 2001, I spoke at invitation-only receptions in the Pennsylvania Convention Center on the opening Sunday of the Philadelphia Flower Show. We hosted similar receptions for donors at the Pittsburgh home of alumna Diane Blanton and the Ford City home of Jim and Lynn Ramage. Through Penn State Extension, we prepared displays for the Philadelphia Flower Show, Ag Progress Days, the Pennsylvania Garden Expo, the Harrisburg Garden Expo, and the Pennsylvania Green Expo. Sometimes these were accompanied by brief presentations from me or Rob Berghage (faculty, Department of Horticulture).

Roger Williams (executive director of the Penn State Alumni Association) and his staff arranged two special receptions as regional Alumni Association events, one at the United States Botanic Garden in June 2006 and another at Longwood Gardens in June 2007. I spoke at these about our plans for an arboretum. I had given a presentation to the Alumni Council in September 2003, and in April of that year, at Rod Kirsch's behest, I also spoke to a large group during the closing celebrations for the "Grand Destiny" campaign. These presentations usually included compelling images from other arboreta and botanic gardens, especially at Big Ten universities. I always mentioned our $10 million challenge, and I invariably got a roomful of amused smiles when I shyly encouraged anyone with interest to speak to me afterward.

My wife, Susie, and I began making purposeful visits to other botanical gardens and arboreta during these years—around thirty altogether, including most of those at Big Ten institutions. I usually spoke with the directors to discuss our project and seek their advice. I met other directors at the annual meetings of the American Public Gardens Association, which I started attending regularly in 2000 and where I made one or two presentations about our plans and progress. These experiences were helpful as we continued with planning and development, and they gave some national exposure to our project.

In early February 2000, I emailed Rod Erickson to suggest that having a name for the arboretum, even a temporary one, would help when I spoke about it before groups. I proposed "The Arboretum at Penn State," which seemed the best way to accommodate the fact that there are small arboreta on campuses other than our flagship campus. Rod liked the name but passed the question of approval to Jan Jacobs, vice president for administration. She thought the name did not need formal approval because the entity it named did not yet exist. We have retained the name, and formal approval probably no longer matters. Admittedly, the name is awkward, and people usually resort to "the Penn State Arboretum" in informal writing. Even official University publications often fail to capitalize the definite article. However, I believe it was the best name we could have chosen in 2000.

In the spring of 2000, I organized a field trip to the arboretum site and issued a public invitation. The mid-May weather was perfect, as it often is at that time of year, and some 100 to 120 people showed up. Paper copies of the Sasaki plan were passed around, and two colleagues—Marc Abrams from the School of Forest Resources and John Ferdinand from the Department of Plant Pathology—discussed research connected with the site. Three experts were on hand to assist with bird and plant identification: Carl Keener and Claude dePamphilis (both from the Department of Biology) and Tim O'Connell (School of Forest Resources). We repeated this with a different set of presenters in May 2001, when 70 people put their names on a sign-in sheet that was circulated.

Whether the attendees at these field trips thought their time was well spent I cannot say. I do know that I felt a little awkward at the artifice of walking through fields and woods, speaking at length about something that was not there and might never be. We stopped organizing field trips after 2001 because there was nothing new to say, at least in that setting. Many years later, we recognized Charles "Skip" Smith in one of the photos taken on the first field trip. Skip Smith gave us the leadership gift for the arboretum in 2007. I did not know him in 2000, but I am all but certain that he did not speak or ask questions. If that field trip helped capture his interest, it was certainly worthwhile despite my doubts at the time.

To cultivate University support, I was on the lookout for research and outreach projects that would demonstrate the value of the tract as an "arboretum." Most of the property was only coarsely programmed in the Sasaki master plan. Of the approximately three hundred acres beyond the Mitchell Tract, some were actively farmed or pastured by the College of Agricultural Sciences, and some were set aside as University Area Joint Authority right-of-way for its wastewater line, but about half of the area was forest or long-abandoned pasture. These lands presented opportunities for long-term research and outreach projects such as those typically found in university arboreta of this size, and I was particularly interested in mission-appropriate installations of merit that could have educational value and special appeal to the public.

I did not solicit project proposals, but a few came my way. The best of these was a proposal from John Skelly (faculty, Department of Plant Pathology) for an Air Quality Learning and Demonstration Center (AQLDC). It was perfect because it added a strong outreach component to some prominent faculty research, it required little area, it integrated perfectly with the arboretum's mission, and Skelly had succeeded in obtaining state and industry funding for the project. Installation of the facility was completed in the summer of 2003.

The AQLDC featured a large LED display of real-time air quality measurements alongside several plant growth plots, some of which were within open-top chambers infused with clean, filtered air. The growth plots dramatically demonstrated how plant development can be severely retarded by regionally ambient (but unnaturally elevated) levels of ozone in the air. The exhibits were impressive and educational, and they helped increase awareness of the consequences of air pollution. However, Skelly eventually retired and left the area, and interest in the AQLDC waned as regional ozone levels declined following changes in the National Ambient Air Quality Standards. Most of the facility was dismantled in 2019, but we retained its roofed pavilion for future use.

Around the time of Skelly's project, in the early 2000s, four of us on the Penn State faculty were collaborating with The American Chestnut Foundation (TACF) on various aspects of its program to breed a blight-resistant American chestnut. Breeding trees is labor-intensive and time-consuming, and TACF's program relied heavily on "citizen-scientist" volunteers. Its Pennsylvania chapter was the largest and most active. The national program had progressed to the point where one or two advanced breeding orchards were about to be established. I was on TACF's national board of directors, and I successfully lobbied the organization to establish one of the breeding orchards in the arboretum and station one of its staff members, Sara Fitzsimmons, at University Park.

A chestnut breeding orchard was inaugurated in the arboretum at a public event in June 2002. A memorandum of understanding between Penn State and TACF was signed, and Bob Steele and the chair of TACF's board of directors planted the first tree. John Oliver, secretary of the Department of Conservation and Natural Resources and a member of TACF, kindly traveled from Harrisburg to participate and provide remarks from the podium. The following day, the *Centre Daily Times* published an article about the event with a photograph. The news was also carried by the Associated Press and on CNN Headline News.

The orchard is on former pastureland in a remote area of the arboretum. Progress toward the goal of a blight-resistant American chestnut is slow because of the many genes that must be captured through breeding. However, the way forward is clear and is still being pursued in the arboretum. TACF now has a branch office at University Park with four employees.

All good arboreta perform a public education function, and they often house excellent research programs. Staffing the arboretum compounded the difficulty of finding the funds to build it, but I believed we could largely avoid having teachers and scientists on the payroll given the presence of five thousand or more full-time faculty employees just across the street, many of whom could be expected to "staff" the arboretum through their everyday scholarly pursuits. The AQLDC and the chestnut breeding orchard helped validate that model and gave us some beneficial publicity in the bargain.

As we endeavored to sustain interest and expectation, I was aided greatly by Kathleen "Kate" Reeder, who became the arboretum's first staff member in July 2000. Funding for her position was drawn from the money allocated by the provost, later supplemented for many years by annual contributions from the Environmental Consortium. Kate's responsibilities shifted and expanded over time, ultimately landing on oversight of publicity, marketing, and space rentals for weddings and other events. With her two degrees in English, she became part of a core group of two or three of us who edited one another's written communications to the public. She remained a dedicated and loyal employee for nearly twenty-three years.

For several of those years, beginning in the autumn of 2000, Kate and I published newsletters with voluntary assistance from Betsie Blumberg, a part-time writer for the National Park Service office in Ferguson Building, where we were also located then. We wrote what we could about an arboretum that did not exist and supplemented that with interesting material related to the arboretum property and arboreta in general. The first two issues, for example, contained articles about the University's well field and wild plants on the arboretum property, ecological restoration work in an arboretum woodlot, our first open house, and the prominent arboreta of two Big Ten schools (Minnesota and Wisconsin). We illustrated the newsletters with line drawings of local plants prepared by Rae Dickson Chambers, a graphic artist in the College of Agricultural Sciences.[1]

The activities described above reflect the general nature of arboretum doings in the years from 2000 to 2007 as we sought to improve our chances of attracting a lead gift. The next few chapters describe several larger projects that were undertaken to directly advance that goal.

The Bellefonte Central Rail Trail

Kate Reeder's first major assignment was to investigate the process of applying for state funding that could be used to convert an abandoned railroad right-of-way[1] on arboretum property into a 1.3-mile bicycle trail. The trail was part of the Sasaki plan, and it seemed a worthwhile project even if the rest of the arboretum were never built. I had gotten enthusiastic support for this idea from both Bill Anderson (assistant vice president for physical plant) and Ron Woodhead, who was director of the Park and Recreation Authority in the Centre Region Council of Governments (COG). Woodhead's cooperation became essential because, as we learned, Penn State would not be eligible for "rails-to-trails" funding unless applications were routed through COG.

In October 2000, following Kate's attendance in August at a state-run workshop on the grant application process, she and I hosted a planning meeting attended by Ron Woodhead, Tom Kurtz (Patton Township manager), Mark Kunkle (Ferguson Township manager), Brian Dempsey and Jude Simpson (Centre Region Bicycle Coalition), and Alan Graefe (faculty, Department of Recreation, Park, and Tourism Management and member of the Faculty and Staff Advisory Committee). The meeting yielded much useful advice. Trish Meek, a transportation planner with COG's Centre Regional Planning Agency, quickly became our collaborator and guide to the intricacies of state funding for local projects. The trail idea rapidly gained momentum over the following weeks as it was presented in a Centre Region Park and Recreation Authority meeting and in public meetings in Patton Township, Ferguson Township, State College Township, and State College Borough.

I had assumed we could get all the funding we needed from a Department of Conservation and

Natural Resources (DCNR) program that supported rail-to-trail conversions. Unfortunately, their guidelines only allowed funding for 50 percent of construction costs, which were estimated to be $426,000. Trish recommended that we seek the balance by applying for federal Transportation Enhancement funds through the Pennsylvania Department of Transportation (PennDOT). One remaining shortfall was that planning expenses, estimated at $73,000, had to be funded locally. For this, we hoped to gain commitments for a cost-sharing arrangement among the University and the four municipalities mentioned above. All three townships touched portions of the proposed trail, and we believed the borough would be interested because of the trail's connection with the McKee Street / Clinton Avenue bike path through Sunset Park. The project management team grew rather large because of the number of offices eventually involved.[2]

With Trish's help, Kate prepared applications to both DCNR and PennDOT. They were submitted in October 2001 with endorsements from local government offices and letters of support from the Centre Region Bike Coalition, Pennsylvania Rails-to-Trails Conservancy, and Centre County Convention and Visitors Bureau. Penn State was the applicant and would manage the project. Also in October, Rick Gilmore and Brian Wimer of the Centre Rails-to-Trails Association began a series of organized trail-clearing work sessions with local bicycling enthusiasts. This was a significant cost-saving measure for the project, because brush encroachment had reduced most of the railroad bed to a narrow path; a part of it, fifty yards or so near the southern end, had no path at all. Here it was possible only to crawl through the tangle of privet, honeysuckle, and multiflora rose (which I did in the company of local officials before the project began). Volunteers had contributed an estimated eight hundred hours of labor by the time we were ready to send plans out for bid.

Gilmore and his colleagues also successfully lobbied each municipality for funding to cover its share of the planning cost. By October 2002, we learned that both proposals to the state had been approved in the full amounts: $213,000 each from DCNR and PennDOT. We had needed buy-in from six government entities to make the project happen, and we got it—a double hat trick. Getting all those pieces to fall into place had seemed like a long shot, so we felt very lucky.

Problems ensued when design costs escalated and construction bids came in higher than expected. The University agreed to contribute $68,000 toward the unplanned expenses, and PennDOT and the Centre County Municipal Planning Organization covered the remainder. The final price tag was $625,000.

These few lines hardly do justice to the hurdles and delays that came our way, and we despaired at times that construction would never begin. But it did, and the trail was formally opened on May 19, 2006. Before we started, Joe Paterno, who lived near the foot of McKee Street, was perhaps the only person who regularly walked the old, overgrown railroad bed. Now it is used by many for recreation and commuting to campus from the Toftrees neighborhood. For a while we hoped the trail could someday be extended all the way to Bellefonte along the old railroad line. However, the right-of-way was ceded to adjacent landowners when the line was abandoned, and a later planning study determined that it would be very difficult to re-create.

The Mitchell Tract Master Plan

As Gary Schultz suggested during the President's Council meeting in April 1999, fundraising and planning for the arboretum required a more detailed plan for the Mitchell Tract than was possible in the Sasaki study. What was needed would probably entail twice the work of the Sasaki plan and encompass only one-sixth of the area. I suspected that a more focused project would yield an entirely different plan for the Mitchell Tract than the extensive parterre arrangement of research and display gardens proposed by Sasaki (appendix C), and a different plan might have more donor appeal. The Sasaki team offered the parterre design as a historic nod to gardens that existed in front of Old Botany Building in the nineteenth century. However, no one alive ever saw those gardens, and they were not a part of common Penn State lore, so I had always been skeptical of the rationale. In addition to a new design (or an affirmation and refinement of the original), we needed a careful financial analysis of expected revenue and operating costs.

By this time, the Mitchell Tract residence (formerly known as Lisnaward) had been renovated to serve as the University president's home, called Schreyer House, and President Spanier wanted to make some changes to the property that could affect the future arboretum. Eliza Pennypacker, still in her temporary assignment as a planner for OPP, suggested that she and I send coordinated requests to Gary to hire a single firm to develop master plans for Schreyer House and the Mitchell Tract. We did so in March 2000, and Gary approved the requests.

This may be a good occasion to address a criticism that came up a few times and may have remained unspoken on other occasions. It was most forcefully brought to my attention by a woman who objected to hiring consultants to plan the arboretum when Penn State has one of the best landscape architecture programs in the country. "Surely," she said, "we have faculty members who could do that work." My response was that perhaps we do, but every design project is different, and every designer brings unique talents, interests, and experiences to projects. We wanted the best designers for *this* project that could be found in a nationwide pool of candidates. I pointed out that our faculty members were free to submit proposals, as they do regularly and successfully for other design projects around the country. And if it were simply a matter of assigning University work to capable faculty members, then we would have our business faculty managing the University's finances, the law faculty handling our legal issues, and the architecture faculty designing our buildings. Penn State does not operate that way, of course, and we treated the arboretum project no less seriously—not as something that could be done as a class project or by a faculty member in his or her spare time.

For the Mitchell Tract plan, we formed a core design team consisting of John Joseph, Dave Zehngut, Neil Korostoff (faculty, Department of Landscape Architecture), Judy Larkin (a landscape architect in the Design and Construction unit of OPP), and me. Judy was our assigned project manager and was new to the arboretum project. She later managed the Bellefonte Central Rail Trail project in 2002, and she remained OPP's project manager for all subsequent arboretum design and construction projects, other than the cultural district plan and art museum, until she retired in 2021. She was always an enthusiastic and diligent advocate for the arboretum, and she has continued her service as a volunteer docent after retirement.

A request for proposals was sent to eight firms in September 2000. Enclosed with the request was a program summary (appendix D) that, with input from the Faculty and Staff Advisory Committee, summarized our vision for this portion of the arboretum. Seven firms responded with proposals, and after interviewing three, we selected Marshall·Tyler·Rausch, LLC (M·T·R), of Pittsburgh, Pennsylvania.[1] Melissa "Missy" Marshall was the lead designer, assisted by Cindy Tyler and Bob Vukich (a Penn State alumnus) and occasionally accompanied by her partner, Geoff Rausch. Dean Runyon of Dean Runyon Associates, Portland, Oregon, was hired as a subcontractor to identify revenue-generating opportunities and prepare a preliminary business plan.

M·T·R made seven visits to campus from April 2001 to the end of that year. These involved workshops and meetings with many groups, most on more than one occasion: the Faculty and Staff Advisory Committee, community leaders and special-interest community groups, selected faculty and staff, OPP facilities specialists, the Arboretum Steering Committee, members of central administration, and the Facilities Planning and Advisory Board. The sessions generated a wealth of ideas and feedback, and Kate's typed notes from each meeting ran to many pages. Missy was searching for garden themes and elements that would appeal to the University and community, and her design was influenced by comments and suggestions we received not only in this project but in later refinements of the Mitchell Tract master plan.

The core design team and stakeholders reviewed the draft plan in December. After some adjustments, the final plan (appendix E) was presented in January 2002 to the Faculty and Staff Advisory Committee and, in a separate meeting, to the administration (Graham Spanier, Rod Erickson, Gary Schultz, Rod Kirsch, Bill Anderson, and Arboretum Steering Committee members). The following day, we presented the plan to a plenary gathering of others who had attended previous meetings. It took several more months to complete a roughly eighty-page narrative, but the plan presented to the president in January, which included the Schreyer House master plan, remained unchanged. Total construction costs were estimated to be $41.4 million for the arboretum portion.

I enjoyed working with Missy. Years before, she had been hired to create plans for Longwood Gardens in her first job after graduating from Harvard's School of Design. Fred Roberts, the director of Longwood, told me in 2002 that he and Missy were "back on speaking terms," but he warned that Missy had "high ideas and strong opinions." I did not mind her assertiveness and valued her smarts, talent, and adaptability in the face of occasional well-founded disagreement from our oversight teams. We got along fine.

The task of the designer involves a certain degree of salesmanship, which is perfectly appropriate. Designers must convince clients they are receiving something of value, and designers usually have a preference among the design alternatives they present. However, I disliked the auteur-like pretentiousness of some we encountered during more than thirty interviews for our various design projects. Exposure to these individuals gave me a greater appreciation of the quiet talents of others who possessed true genius.

Missy Marshall was one of those individuals, and Emmanuel Didier (whose work for us is described later in this book) was another. During his first meeting with our oversight group, Emmanuel asked each of us to rate the appeal of twenty or thirty photographic vignettes taped on the walls of the room, all from other gardens and probably pulled from the internet. He gathered our individual rankings and listened to comments, and he returned the next day with sketches that bore no obvious relationship to the photographs but were immediately compelling to us and established a direction for our garden design. Were we guiding him, or was he guiding us? There was no answer, but the result was a garden design that had our unanimous support.

Missy's design approach, as I understood it, began with identifying and evaluating site constraints, candidate garden elements, core circulation patterns, and sight lines and views. She displayed an intuitive grasp of the likely visitor experience and made selective adjustments to the design as we continued the process of gathering feedback from stakeholders and the core design team. Her master plan for the Mitchell Tract laid the foundation for the arboretum's current popularity, and we were fortunate that she was our consultant at this early stage. The master plan and the first construction, completed in 2009, represent her main contributions, and many elements of the master plan have been modified over the last twenty years. However, her influence pervades almost everything that is there now. Even our last project, the Pollinator and Bird Garden designed by Emmanuel Didier, owes much to a preliminary design developed by Missy.

As we worked on the Mitchell Tract master plan, it became clear that design decisions should be guided by more than ad hoc judgments that are necessarily subject to circumstance and committee membership. One member of the Faculty and Staff Advisory Committee, Peter Everett (faculty, Department of Marketing), suggested writing a set of design principles to guide future design work.

After some discussion, Eliza Pennypacker agreed to lead this effort, and I created an ad hoc committee following her membership recommendations.[2] Its responsibility was to develop (1) guidance for resolving design issues and evaluating project proposals not addressed by existing plans and (2) guidelines for consultants hired to develop new plans.

The committee completed its work in May 2004. After review by Ford Stryker (associate vice president for physical plant), Gary Schultz, and the Facilities Planning and Advisory Board, the guidelines were approved with minor changes in April 2005 (see appendix F). I was assured that they were incorporated into OPP policy and distributed as background material for subsequent design projects.

Following one of the recommendations in the guidelines, I created a standing committee on arboretum design in November.[3] Members of the Arboretum Design Committee contributed to every significant arboretum planning project during their tenures, and the committee was occasionally consulted to provide advice on special issues and projects. Typically, I entrusted oversight of the guidelines to the professionals on the Arboretum Design Committee, and two or more of them served on every project oversight team.

However, I was always very mindful of one guideline on life-cycle costs and authenticity. Given that usage and time always take a toll and that maintenance and repairs will someday be at the discretion of others, it seemed important that design decisions should ensure a lasting appearance of underlying quality. By analogy, if the arboretum were a winter coat that might eventually become scuffed and worn at the cuffs and collar, we wanted it to be fashioned from fine leather rather than, say, vinyl leatherette. The idea was to ensure that the arboretum would always be worth treasuring and, if necessary, restoring. Like medieval cathedrals that retain their magnificence even if the stone steps are deeply worn and the stained glass discolored, we wanted an arboretum with a "great deal of ruin" in it. Admittedly, such a goal may be more achievable with a cathedral than a garden of living plants, but it seemed worth the reach. Of course, we also hoped to create something so beautiful that it would never be allowed to fall into neglect.

Design guidelines or not, each of us centrally involved in planning and building the arboretum came to internalize similar ambitions for what we were creating. Gordon Turow, the OPP landscape architect who was so important to our planning work, once told an interviewer, "Every near-term decision that we make is anchored in a long-term perspective. We will never compromise our design standards, our attention to detail, or our quality. Permanence will never be a question—each move that we make in the arboretum will be a legacy move." Perhaps because we intended the arboretum to become a unique addition to campus, this project seemed to demand an exceptional level of quality. President Thomas's dictum to "do it right" had become our own.

The Jump-Start Plan

Even as the future existence of the arboretum remained very uncertain in the early to mid-2000s, several campus buildings were located and planned in ways to take advantage of it. The Lewis Katz Building, completed in 2007, was sited next to the Mitchell Tract at the northeastern corner of Bigler and Park because, I was told, the law school dean liked the adjacency to the future arboretum. The Katz Building displaced a "flower trials" garden, which annually featured new flower cultivars in trials operated by the Department of Horticulture. The impending loss of this garden partially catalyzed the planning discussions described below.

Opposite the Mitchell Tract, across Park Avenue, was a large student parking lot (Lot 80) at the southwestern corner of Bigler and Park. The 1999 University Park campus master plan recommended that this space would be better utilized for academic buildings, and the plan suggested a design that would complement a future arboretum on the other side of Park Avenue. In this design, buildings were arranged behind a large, semicircular lawn facing Park Avenue (and the future arboretum) while avoiding a central sight line to the arboretum from a walkway that extended from Curtin Road, next to where outdoor seating for the Berkey Creamery is now located. The proposed arrangement would afford views of the arboretum from the buildings, the walkway, and even Curtin Road. Those buildings became Business (2005) and Forest Resources (2006).

Now, twenty-five years later, the old Lot 80 site has been redeveloped exactly as envisioned in 1999. However, the original design intentions were nearly scuttled during an early planning meeting for the Business Building. The meeting's attendees were Graham Spanier, Rod Erickson, Gary Schultz, Rod Kirsch, Bill Anderson, Gordon Turow, Judy

Olian (dean, Smeal College of Business), and Robert A. M. Stern (the building's architect). As described to me by a witness, Stern proposed locating the Business Building in the *middle* of the block, at the apex of the semicircular lawn. This location would have prominently centered Stern's building in the available space, blocking the sight line from Curtin Road and limiting the size of future buildings on either side.

This was a direct challenge to the University's master plan—and the sort of challenge that is not always beaten back, whether from indifference or out of deference to the architect. Stern was one of the world's most renowned architects, and the charmingly formidable Dean Olian was raising large amounts of money for her costly building. At this point, Bill Anderson whispered to Gordon, "You had better do something NOW!" Never the central figure in meetings like this, Gordon was always prepared to put his hands on the wheel and quietly reset a wayward discussion with an incisive, convincing summation and commentary. On this occasion, he picked his opportunity, and Anderson backed him up. The Business Building did not get the location its architect wanted, and the sight line to the arboretum was preserved.[1]

As the planned arboretum became a factor in the location and design of new academic buildings, I grew more hopeful about its inevitability: these decisions only strengthened the general belief that building the arboretum would be a very good thing. The Business and Forest Resources building projects were on the horizon when Bob Steele and I met with Gary Schultz and Rod Kirsch in October 2003 to discuss arboretum funding. One topic of our meeting was a proposal to pursue a multimillion-dollar legislative appropriation for Pennsylvania arboreta, an idea that the directors of several gardens and arboreta[2] had discussed with Bob and me since 2000. Gary had been open-minded about the proposal—but cautious. A concrete proposal from the principals was now in hand, to be endorsed by Penn State or not, and in a subsequent meeting with Rich DiEugenio (director, Office of Government Affairs) and Mike DiRaimo (director of governmental relations), Gary made the decision to scuttle Penn State's participation. His concern was that any money we might get in this way would be taken from the University's existing request to the state for $40 million to fund capital projects. The legislative initiative was ultimately successful for the other arboreta and gardens.

The other topic was the possibility of starting the arboretum with fundraising in $1 million to $2 million increments while we sought an eight-figure donor for an education center / administrative building. The approach seemed feasible. (Indeed, its feasibility was later demonstrated by our success in adding seven-figure capital projects, each funded by gifts, to the initial arboretum construction.) However, it required Graham's approval because it would represent a break from the original plan. Rod made a commitment to hire a development director for the arboretum if the plan received approval.

This became the "jump-start plan." When we met, the location for the Katz Building had not yet been chosen. However, a major driver behind the plan was the desirability and, later, the necessity (after the Katz site was chosen) of relocating the flower trials garden. At Gary's direction, Gordon Turow, Bill Anderson, and I developed a scheme that became the subject of three meetings with Gary and Rod Kirsch through mid-February 2004. It called for central administration to contribute $3.8 million to develop infrastructure (utilities, grading, parking) and move the flower trials garden. The relocated garden would be the initial piece of the Demonstration Gardens as programmed in the 2002 Mitchell Tract master plan. The final part of our plan was a portfolio of fundraising opportunities for other elements of the master plan, including the Pond and Fountain, Four Seasons Garden, Wetland and Aquatic Garden, and March Bowl.

We met with Graham in April 2004 and left with a commitment of $3.8 million from the central administration for infrastructure if it could be coupled with one of three gift options: (1) a $10 million donation to fund the master plan's Education Center, (2) a $10 million gift to fund the Conservatory and two gardens, or (3) a minimum of $4 million in gifts bundled from four or five donors to fund the front gardens. Rod thought the third option would be challenging to pull off, and the first two still left us needing a $10 million donor. However, Graham's commitment of $3.8 million for infrastructure was a positive development.

Rod was still talking with Peter Tombros about making a large gift to the arboretum, which was a major reason for the meeting with Graham regarding options. Eventually, in early 2005, Peter did make a very generous offer. Unfortunately, the gift would be distributed over many years, and its present value in 2005 would not meet our needs. We regrouped with a counteroffer under option 3 that was priced for a construction start in 2008. However, Peter's interest was in the Conservatory, and eventually, in 2006, new cost estimates for that feature set us even further apart.

The jump-start plan and the impending construction of the Katz Building made it necessary to decide how to incorporate the flower trials garden into the arboretum. The University signed a contract with Sasaki Associates in September 2005 to devise a plan. Throughout the fall and early winter, we received verbal and written input from Rich Marini (head, Department of Horticulture) regarding their needs vis-à-vis the flower trials garden and how to accommodate them in the Demonstration Gardens. I sought similar feedback from all of the other department heads in the College of Agricultural Sciences on the assumption that the Demonstration Gardens could be a vehicle for wider academic collaborations.

Unfortunately, what we learned gave me little confidence that the Demonstration Gardens would work as I had expected. Marini reported that the flower trials garden was a burden to his department, requiring an annual subsidy of about $30,000, and his faculty believed that the gardens had limited academic value. He concluded that there was no reason to salvage more than 10 percent of the existing garden. Of the other department heads, several suggested ideas for demonstration plantings, but it was unclear how (or even whether) their faculties would use them for teaching and research or for Extension programs. I began to believe this approach would simply lead us to a set piece of agriculture-themed plantings of no special merit, not a dynamic venue for faculty engagement with the public.

By early spring 2006, Sasaki's emerging design for the Demonstration Gardens had increasingly seemed uninspired and disconnected from M·T·R's master plan. Also, the plan for the Katz Building had been completed, and it became obvious that the building would render some elements of the Mitchell Tract master plan obsolete. Gordon Turow, Paul Wangsness, and I met in March 2006 to discuss our next steps. We agreed to part ways with Sasaki and decided to seek the services of Missy Marshall. In April, we contracted with her firm to update the 2002 Mitchell Tract master plan and recommend what to do with the flower trials garden.

In this new project, the most important changes to the 2002 plan (appendix E) began as bullet points agreed upon at a charette conducted by Missy and her crew in May 2006.[3] The major components of the original plan were retained, except for the elimination of a large pond near Park Avenue. However, many things were rearranged in response to the construction of the Katz Building and the Business and Forest Resources Buildings across the street. The plan that eventually grew out of this meeting (appendix G) became the genesis for the Mitchell Tract gardens as they appear now.

As meetings and refinements continued, Missy cautioned that placing the flower trials garden in the arboretum would not create the visual impact needed to attract donors. This advice, together with the earlier feedback from the Department of Horticulture, led the College of Agricultural Sciences to conclude that the flower trials garden at University Park was completely expendable.[4] The final version of the master plan was completed in September 2006 and included a phasing plan for incremental construction. Total build-out of the new plan was estimated to cost $34.6 million.[5]

Our Witness Tree

As early as 1999, I began getting suggestions that we simply plant some trees and put up an informal arboretum sign. The notion was that a collection of trees planted at minimal cost might secure the space as an "arboretum" even if we never got the gift to fund our grander vision. Free trees were often available as surplus from research projects, and donations could be obtained from other arboreta. Countering these suggestions was President Thomas's admonition to "do it right" or not at all.

In addition, I had come to believe that it might be easier to secure a large gift to *create* an arboretum than a smaller gift to *improve* one we already had; our best chance to have a great arboretum was before we had anything. Making a transformational gift—for a scholarship, a new building, a novel scholarly center—has an innate appeal that other gifts do not. I realized this after visiting several older arboreta that had started with little investment. One of these (the Nichols Arboretum at the University of Michigan) was basically a century-old neighborhood park. It had a local following that liked the arboretum as it was and did not seem to regret what it was not. Even if the Nichols suddenly developed new and expensive ambitions, its prospects for an eight-figure gift would be limited by contentment with what was already there. It was not yet clear whether enthusiasm for our project was strong enough to deliver the gift we needed, but whatever enthusiasm there was would certainly dissipate if we started on the cheap with a few saplings and a sign.

This was my point of view when we received a call in September 2001 from George Biemesderfer, owner of Green Acres Nursery in Lititz, Pennsylvania. George told Kate that he wanted to donate a tree to the arboretum. That was not our first such offer, and I was prepared to decline when I returned the call. However, what George had in mind gave me pause. His desire was to donate a "very large" tree in honor of Charles Hosler and his wife, Anna Rosa Hosler. Charlie Hosler was a retired University vice president[1] who, as a young faculty member in 1950, had helped his younger and somewhat bewildered cousin George settle in as a Penn State freshman. George wanted to honor his cousin and hoped that the tree, if sufficiently large and prominent, would serve as a beacon and inspiration for other donors. His nursery was experienced in the unusual specialty of moving and planting very large trees.

What convinced me to accept George's offer was the suggestion from a faculty colleague, Henry Gerhold, that the arboretum should have a "witness tree"—a ceremonial planting akin to a building's cornerstone. The tree would "witness" the creation and growth of the arboretum and, of course, it had to be there at the beginning. Henry's inspiration came from a white oak on the property of the Donegal Presbyterian Church in Lancaster County. It was locally called the Witness Tree because a colonial congregation had gathered beneath it in 1777 to commit themselves to the American Revolution. (British troops were just then entering Pennsylvania, and not everyone in America was unhappy with that fact.) The tree was large in 1777, and it lived for another 218 years. As I saw it, a witness tree should be a notable tree already present when the arboretum was inaugurated, and it would be wonderful if it were a gift of special significance and strategically located within the landscape of the future gardens.

George made a list of ten candidate trees on his nursery property, and Jim Sellmer, Derek Kalp[2] (landscape architect, Campus Planning and Design), and I drove to Lititz in October 2001 to inspect the trees. We wanted a tree that would live for centuries and selected a white oak, many of which are in the arboretum's Hartley Wood and as old as 360 years (appendix B). Jim, a specialist in ornamental trees, volunteered to be an adviser and liaison for the operation. Jeff Dice, whom I have mentioned previously, offered to prepare the arboretum site and provide a crane when the time came to shift the tree from George's truck.

White oaks, especially large ones, are notoriously difficult to transplant successfully. Preparations to move our witness tree began in November 2001, when George trenched around the tree to encourage the development of small roots close to the trunk. There was more work required over the next three years, and finally, a test dig in January 2005 indicated to George that the tree was ready to move.

By this time, there had been no significant press coverage of the arboretum for three years or more, and I was deeply worried that prospects for our lead gift were fading from a sense that the project was moribund. The witness tree became ready to plant at an opportune time; we needed a reason for people to believe the project was still alive. Privately, though, the witness tree installation felt like a "Hail Mary pass"—a final act of desperation. I had no more ideas to reinvigorate the enthusiasm and expectation that had been so prevalent in 1999.

We planted our witness tree, the Hosler Oak, on March 21, 2005, in the open field of the Mitchell Tract bordering Park Avenue, an area used only for football parking. We had purchased eight shovels from a hardware store and painted the blades gold. With these, the tree was ceremonially "planted" by George, Charlie, and me, accompanied by Rod Erickson, Gary Schultz, Rod Kirsch, Bob Steele, and Eric Barron (the new dean of the College of

Earth and Mineral Sciences). All the others present then took a turn with a shovel to participate in the moment. Sixteen years later, as president of the University, Eric admitted to me his grave doubts about the arboretum in 2005 and said he told his wife, Molly, after the event that the Hosler Oak would probably be the loneliest tree on campus.

At thirty-three feet tall, with a trunk diameter of fourteen inches at a foot above the ground, and weighing twenty-nine thousand pounds with its root ball, the Hosler Oak was the largest tree planted on campus since two elms were placed at the front corners of Old Main in 1933. According to George's records, the tree was between thirty-seven and thirty-nine years old. After planting, it required special care and attention, and Jim Sellmer took a personal and professional interest in the project. With the assistance of one of his students, Dave Despot, Jim installed soil moisture probes in the planting hole and monitored those readings throughout the 2005 growing season. Periodic measurements of leaf water potential, taken before dawn in an elaborate procedure, provided direct readings of plant water status and helped calibrate and interpret data from the soil moisture probes. Irrigation lines were installed and fed by a water tank that Jim and Dave periodically refilled from a distant water source on campus. Between planting and late autumn, they contributed an astonishing 168 man-hours to keeping the tree healthy in 2005.

Because there was always a possibility that weather conditions might force us to postpone the planting date, we planned a more public event for September 28, 2005. Kate arranged the catering and publicity for this occasion with the help of Bob Steele's staff and Barb Meeker in Penn State's Office of Campus and Community Affairs. We rented a tent. I asked the president's administrative assistant, Carolyn Dolbin, about getting the event on Graham's calendar, and Graham agreed to deliver the opening and closing remarks. Paula Ammerman, director of the Office of the Board of Trustees, arranged for Anne Riley to speak as a member of the board. I knew her and was aware of her enthusiasm for the arboretum plan. Bob Steele and Charlie Hosler planned to speak, as did I.

We have no record of how many attended, but based on memory and photographs, about 150 people turned out. The day was beautiful, and everyone seemed to enjoy the occasion. The only disappointment was when Graham, perhaps disturbed by the very public nature of the occasion, took me aside to complain about the "pretense of having an arboretum." I did not see it that way and explained why. I understand in retrospect why some could have interpreted the event as the inauguration of something greater, but I did not intend it to be a "stunt," as Graham's comment suggested.

I sometimes give First Year Seminar tours to students. I like to pause at the Hosler Oak to explain that it is the arboretum's "witness tree" and could live for another three hundred years. I tell them to remember their visit because the tree will still be there, only much larger, when they return for their fiftieth-year reunions. I'm sure some will remember.

The End . . . and Then a Beginning

The year following the Hosler Oak planting was marked by two developments with opposite effects. On May 1, 2006, I was summoned to a meeting in the president's office with Graham Spanier, Rod Erickson, Gary Schultz, Rod Kirsch, and Bob Steele. I assumed that the main purpose of the meeting was to discuss the place of the arboretum, if any, in a University capital campaign that was under preparation. However, Graham's interest was in another matter that he quickly raised: criticism of my actions to promote the arboretum. This was clearly precipitated by the Hosler Oak planting, but in fairness, the witness tree was only the most recent occasion when Kate and I did something that might have prompted board members or others to ask awkward questions about progress with the arboretum.

Graham asked that I step away from the microphone, as it were, and "mothball" all publicity until the money came in. In effect, I was dismissed from active duty as director. The others took little part in the meeting, and it was only later that I realized Graham's words were aimed at them as well as at me, as if to say, indirectly, "Stop enabling him!" They had been, but mostly in the sense that an occasional question from me would always receive a timely and helpful reply, and their guidance and assistance were invariably proper. The meeting was distressing, and I was certain that the project had received a lethal blow. Ironically, it happened just as we were quietly celebrating the completion of the Bellefonte Central Rail Trail project.

It now seemed very unlikely that a donor would materialize if everything went quiet almost nine years after the arboretum project became public knowledge.[1] As Bob and I left the room to return to Ag Hill, Rod Erickson spoke quietly and apologetically: "You realize, don't you, that I will have to pull your funding?" This meant letting Kate Reeder go and losing our small operating fund. On our walk back, Bob said he could keep us afloat for a while, which I interpreted to mean a few months after my allocation from Rod ran out on June 30.[2]

What I did not know at the time, but Rod Kirsch and perhaps Graham Spanier did, was that a local businessman had telephoned the Office of University Development a few weeks earlier to inquire about the arboretum, stating that he might be willing to make a $10 million gift to the project. In June and again in August, the man returned in person with his son David to discuss an arboretum donation with Rod Kirsch and Dave Lieb, associate vice president for development. Rod alerted me to the appearance of a potential donor after the June meeting, when the possibility of a leadership gift from this person began to look very strong.

The potential donor was Charles "Skip" Smith, who wished to honor his father with a naming gift. His father was H. O. Smith, a 1920 graduate of Penn State who settled in State College as an architect and real estate entrepreneur. Skip revealed in the August meeting that he was the anonymous client whose financial adviser had inquired in 2000 about donating to the arboretum. Skip was known to Gary Schultz and others in Old Main through real estate transactions, but he had not been a notable donor to the University. Gary had made a comment in October 1999, during our meeting about the arboretum directorship, to the effect that Skip Smith (a name unfamiliar to me at the time) was interested in the arboretum. As I mentioned some pages back, Skip attended our first open house event in May 2000. He had clearly been considering a gift to the arboretum for many years, and he showed press clippings to Rod and Dave that he had saved during that time. The Hosler Oak did not awaken Skip's interest in donating to the arboretum, but I believe it encouraged him to proceed after the project had appeared inactive for several years. (There was little awareness of the rail trail work outside of the cycling community.)

Of course, Skip wanted to know what we would do with a $10 million gift and how his father would be recognized. We had previously regarded the Education Center as the primary target for initial construction, but its projected cost had escalated in Missy's latest estimate. I proposed "botanic gardens" as our best alternative naming opportunity and perhaps the most attractive thing we could offer a donor. This required some more study because we had no area in the plan called "botanic gardens."

Fortuitously, these discussions with Skip dovetailed nicely with Missy's work on revising the Mitchell Tract master plan and its cost estimates. In August 2006, before the third meeting with Skip, Gary asked OPP to create a plan to build part of the botanic gardens for $10 million and to assist me in preparing a business plan that reflected revenue generation and operating expenses. At Ford Stryker's request, Mark Bodenschatz (director, Design and Construction, OPP) and I collaborated on the numbers using data from Missy as our raw material. Ultimately, we created a spreadsheet showing projected costs for a phased build-out of the entire Mitchell Tract master plan.

Around this time, a chance conversation with Jeff Dice, who supervised OPP's grounds maintenance operations, led to a significant breakthrough in our cost estimates. The 1999 Sasaki and 2002 M·T·R plans included a maintenance center whose estimated cost was in the low millions of dollars. With added costs for equipment, the capital outlay for arboretum maintenance had always seemed an almost insurmountable hurdle for fundraising. Bucket trucks and pole barns do not have high donor appeal.

I knew Jeff well through our mutual service on the Campus Tree Commission, and on this occasion we sat next to each other at a campus seminar. He asked about the arboretum project, and I described my concern about finding the money for an in-house maintenance operation. Jeff suggested we contract our maintenance work to OPP, as Athletics and a few other campus units do. I was not aware of such arrangements, and I suppose it had never occurred to anyone who knew about them, other than Jeff, to suggest this alternative. Assuming that we would be charged actual costs for this service, contracting with OPP would essentially amount to replacing a huge (to us) upfront capital investment with the more manageable costs of annual capital depreciation and overhead. The labor cost would be a wash.

Jeff was innovative and had high standards, and I knew the arboretum would be well maintained under his care.[3] For Jeff's part, he believed his crews would benefit from working in the arboretum and that the partnership would elevate the quality of landscaping across the entire campus.[4] A critical element of the proposed relationship was that the arboretum staff, while not directly supervising OPP personnel, would oversee the work to ensure that the quality of maintenance met the standards appropriate for a botanic garden.

Handling arboretum maintenance through an OPP contract opened up opportunities for more gardens and greater visitor appeal with a $10 million gift. Still, we needed to determine how much of Skip's gift to set aside to support operations. Gary cautioned me not to rely on Dean Runyon's revenue projections from the 2002 Mitchell Tract project. However, we knew there would be *some* revenue from event rentals and some income from undesignated donations. Gary and I agreed to set aside about half of the $10 million for the endowment. In later years, we recalculated the endowment allocation needed to support each new capital project that came along, always with more reliable projections than we had in 2006, and it consistently worked out to around 50 percent of the total gift. I never sensed that Skip or other donors begrudged this use of their money. In fact, I believe they appreciated that we were providing for the perpetual stewardship of their gifts.

Gary was correct. Runyon's revenue projections never materialized, and earned income from event rentals and other sources falls far short of the arboretum's expenses. Runyon assumed that we would have an entrance fee, which is usually a significant revenue source for public gardens, but admission to the arboretum is free.[5] Although students use the gardens heavily, the arboretum's annual budget is augmented neither by student tuition and fees nor by state appropriations to the University. Endowment income covers most of the cost of operating the arboretum. My only disappointment about revenue is that few visitors use the on-site donation box. We never expected donation-box revenue to be an important part of the budget, but I am sorry that more visitors do not "pay it forward," even in a small way, after enjoying the generosity of others. On the other hand, I am sure that most people do not know the arboretum was built mostly with philanthropy.

By September, Mark Bodenschatz and I, working closely with Missy, finished drafting a construction scheme that looked achievable with a $10 million gift (less allocation to endowment) plus Graham's 2004 commitment of funding for infrastructure and some additional fundraising around identified naming opportunities. We judged it to be enough to create an attractive garden that would have significant visual "presence" from Park Avenue: an Event Lawn and Overlook Pavilion, most of the master plan's walled gardens, a fountain, a parking lot, part of the Demonstration Garden, and landscaping throughout the area between Park Avenue and the walled gardens.[6]

Using our numbers, University Development prepared a proposal in December 2006 to name and partially fund the H. O. Smith Botanic Gardens, now circumscribed within the revised Mitchell Tract master plan (shown in appendix H). Skip met with Rod and Dave no fewer than eight times from June 2006 to April 2007, sometimes accompanied by his son David. Skip had questions about how his gift would be spent and managed, and he was curious about detailed aspects of the arboretum plan. After each meeting, Dave and Rod asked me to provide answers, which I did to the best of my ability. Sometimes, the only possible answer was in the vein of "trust us" or "we'll do the best we can," which he accepted.

I met Skip for the first time on April 26, 2007, in the Alumni Lounge of Old Main, to mark his signing a pledge agreement to donate $10 million over five years. It was a chilly, rainy day, but my gratitude and happiness were undampened. The arboretum project would succeed, bringing ninety-three years of on-and-off effort to fruition! I emailed the news to Bob Steele and Kate Reeder in the evening, stressing the need to keep the information confidential until the Board of Trustees met. Bob had been kept fully informed by Rod Kirsch throughout the discussions with Skip, so he knew what was going to happen that day. Rod asked that I allow the administration to inform the rest of the Arboretum Steering Committee. With his permission, I sent the news to the Faculty and Staff Advisory Committee a few days before the Board of Trustees approved the gift agreement and name on May 18. The University sent out a news release that same day.[7]

The H. O. Smith Botanic Gardens

We started immediately on the Phase I project.[1] Six days after Skip signed the pledge agreement, Ford Stryker issued a special-funding budget request for final design and construction, and he alerted our OPP project manager, Judy Larkin, to begin work. By mid-May 2007, Dave Zehngut had developed a letter of request for proposals and issued invitations to thirteen firms. The deadline for proposals was in June.

Also in May, Gary Schultz circulated a memo summarizing the expected project costs as $15.5 million, comprising $7 million for design and construction and the balance for an operating endowment. Skip's gift of $10 million and a University contribution of $3.1 million left a shortfall of just over $2.4 million. Gary pointed out that the shortfall must be addressed by means of operating income, budget reductions, or a temporary subsidy. We also expected to make up some of the shortfall with additional fundraising. (In the end, although the final cost of design and construction was higher, at $7.3 million, we made ends meet through fundraising and a small subsidy for operations from the provost.)

Only three firms submitted proposals. This was disappointing but perhaps understandable. Some agencies probably assumed that M·T·R (Missy Marshall and crew) would have a strong advantage, given their authorship of the preliminary design. Others may have been dissuaded if they thought their creative contributions would be limited to coloring in a scene that M·T·R had already sketched. However, after interviewing the three firms, we felt certain that M·T·R would be an excellent partner in carrying Phase I through to completion. As M·T·R was strictly a landscape architecture firm, their proposal included subcontracts for engineering, cost estimating, and design of the all-important Overlook Pavilion. For the last, M·T·R partnered with the San Antonio firm of Overland Partners, the architects for Penn State's award-winning Stuckeman Building. The proposed schedule called for design completion by November 2007, construction documents by February 2008, and construction completion by April 2009. The project would encompass improvements on 29.5 acres of the Mitchell Tract bordering Park Avenue.

By August 2007, the Faculty and Staff Advisory Committee became a Phase I oversight team, as mentioned earlier, and we had our first design workshop that month with Missy and her team. We accomplished much in that workshop, but there was a serious disconnect. Missy came to the meeting expecting our guidance on plant selection, and we were unprepared. Missy demonstrated an encyclopedic knowledge about plants. However, she said in her June interview that she hoped to draw on us for plant material suggestions and recommended that we develop a collections policy.[2]

Missy's expectation was reasonable—this was to be *our* arboretum, after all—but I had set aside her recommendation about preparing a collections policy because it would have been impossible to do, necessarily by committee, over the few summer weeks available before our first workshop. Missy and I talked afterward, and we went to work on the Penn State side to identify thematic directions for selecting plants.

One point of discussion among the oversight team was the use of cultivated varieties (cultivars) from the horticultural trade. The horticulturists among us favored cultivars, and the botanists insisted that a botanic garden should represent nature, not the artifice of cultivars developed for novelty or appeal. We reached a compromise when it was pointed out that some cultivars are simply very nice, vigorous representatives of their species. Using such cultivars in a botanic garden is as natural as would be, say, using images of young athletes to represent humankind to an extraterrestrial civilization. Everyone acknowledged that beauty enhances the appreciation of nature, even in a garden that intends to offer more than mere horticultural display.

Botanic gardens in the past were used to study herbal medicine and economic botany, and later, plant classification and evolutionary relationships. It was once common, for example, for gardens and arboreta to physically arrange their collections to illustrate then-conventional notions of the "natural order." These purposes (if not the topics) have become less important, and botanic gardens today serve many functions.[3] However, no function is so thematically comprehensive as to dictate the composition and arrangement of collections, as was often the case in the past.

Our team settled on a theme of "vegetation linkages in space and time," a concept that in its broadest sense can connect plant systematics and phytogeography (traditional botanic garden themes) to topics in evolutionary biology, anthropology, climatology, geology, human culture, and ultimately nature and our place within it. Rather than serving as a template for selecting and arranging plants, it is a simple heuristic device for guiding decisions about acquisitions and reimagining how plant collections can serve academic purposes.

The theme is especially appropriate given the location of our arboretum, which is rather uniquely situated at or near the southern range limits of many northern plant species, the northern limits of many southern species, and the eastern limits of many midwestern species.

Furthermore, the small region surrounding us is nearly unique within the United States for having been continuously vegetated for about four hundred million years; other parts of the continent were under water or ice from time to time during the history of land plants. We never explicitly promoted this phytogeographic individuality, and it does not pervade the design. However, it informed many of our choices of species to feature and inspired some of the named arboretum features, such as the Grove of the Ancients,[4] Tropical Grove,[5] and parts of the Children's Garden.

I sent a four-page draft of our ideas to Missy a couple of weeks after our August meeting. A few days later, she responded with an outline, an executive summary, and bubble diagrams. By early September, Missy had completed minor revisions to the master plan (appendix I), and she returned to campus to focus on details in a three-day design charette. Planning Phase I was well underway and on sound footing! For many of us, the following two years were very busy with arboretum matters.

The oversight team was large (twenty-one members) and perhaps a bit challenging for Missy. I never considered it unwieldy, though, and we always had less than full attendance because of scheduling conflicts. We met biweekly throughout the autumn of 2007 to discuss and respond to Missy's latest design developments, sometimes with her present in person or on speakerphone. The team worked well, and members were generous with their ideas and suggestions. Kate Reeder's typed (and single-spaced) meeting summaries were usually five pages long or more. Many members stayed with us to help plan later additions to the arboretum. One of the most regular and active participants was Andy Stephenson (faculty, Department of Biology), who participated in nearly all of the projects until the completion of the Pollinator and Bird Garden in 2021. He and his wife, Sue, made a large gift for that garden, and Andy continued to contribute as an arboretum volunteer after his retirement around the same time.

A prominent design feature was a walled garden area partitioned into thematic rooms: three rooms plus the Bamboo Allée in Phase I, with four rooms planned for the full build-out of the master plan. The walls were meant to serve two purposes: first, to lend some "presence" to the arboretum as viewed from Park Avenue, with the promise of something unseen just within, and second, to give visitors inside a feeling of intimacy and refuge from campus activity just across Park Avenue. Missy designed the walls to be seven feet high to achieve these purposes.

In a late November briefing, Graham worried that the walled gardens in what was then an open field would resemble "Fort Apache." Before approving the design, he asked OPP to build a plywood mock-up of a section of the wall nearest Park Avenue. This was done just before the holidays, and the plan was approved as drawn. Some of us complained about this extra step because we had already scrutinized the plans with critical eyes, but I was sympathetic with his desire to ensure that things under his purview were done "right." Apple Inc. surely would have been far less successful without Steve Jobs's famously obsessive concern for the look and feel of its products. This attention to detail was one of Graham's strengths. I believe he was also responsible for increasing the budget for campus landscaping during his tenure, and that provided an indirect benefit to the arboretum.

One of our team members, Joyce Robinson, a curator at the Palmer Museum of Art, called me in October 2007 about a proposal to create a stone plaza depicting the Spring Creek watershed, which drains the local landscape. The proposal came from a local artist, Stacy Levy. We were already discussing a watershed theme in the master plan because 85 percent of the University's water is drawn from wells on the property, and Stacy's proposal was readily approved at the next oversight team meeting. By December, a local philanthropist, Don Hamer, agreed to fund the project as a gift to the College of Arts and Architecture. Missy said she could find a place for it at the Overlook Pavilion, which lies on the divide of two small watersheds that feed the Big Hollow aquifer and ultimately Spring Creek. The watershed map (Ridge and Valley Sculpture) has been one of the arboretum's most distinctive and visited features.

Penn State held a groundbreaking ceremony for the arboretum on November 2, 2007.[6] The outline of the future Strolling Garden walkway was painted on the bare turf to give attendees a sense of how the plan would fit into the landscape. As intended in 2005, the Hosler Oak, our witness tree, stood just beside the route of the walkway. About 350 people were present.

Preliminary work to install utilities and stockpile topsoil commenced in November, while the preparation of construction documents continued into 2008. The oversight team met only once more, in February, but we were active throughout the winter with email discussions about the planting design and the evolving plant list. The end of April was the deadline for bids, and the lowest bid, from Leonard S. Fiore, Inc., was very close to our budgeted cost. The quality of plant material and its installation were very important to us, and we were pleased that Scott's Landscaping, a local firm, was Fiore's subcontractor for landscape work. They had done good work on previous campus projects.

Construction under the Fiore contract began in July 2008. In November 2007, from a conference room on the fourth floor of the Forest Resources Building, I took photos with a digital camera aimed at fixed points within the construction site. This became a weekly activity as the pace of construction picked up. After it was finished, Sara Fitzsimmons—a tech-savvy graduate student—helpfully stitched them into registered and aligned panoramas and a time-lapse video. The entire site was also visible from my third-floor office, and I watched and then photographed the first tree (after the Hosler Oak) being planted on March 31, 2009.

The construction project was unexpectedly demanding of my time. There were regular job conferences with the contractors, and emails and phone calls required almost daily attention, often necessitating trips to the job site. Several of us (Judy Larkin, Dave Zehngut, Gordon Turow, and I) met often at the site to discuss the paint tint for the metal parts of the Overlook Pavilion, the color and texture of the stucco on the garden walls, the color and surface treatment of the exposed aggregate walkways, and other small matters, and we frequently asked for changes after reviewing mockups. I sensed that this level of attention was more than usual in a construction project, especially from so many of us, but we all considered this project special.

In early April 2009, Gordon Turow, Dave Lieb, Patrick Williams (the arboretum's new development director), and I gave Skip a behind-the-fence tour of the construction site. Our tour circled the still-unseeded Event Lawn along the Strolling Garden walkway.[7] We paused to talk near the end, and somehow—I have forgotten why—the Old Main bell entered the conversation. At the time, Gordon was working on a class gift project to move the bell to a display on the nearby lawn. Skip mentioned that his uncle or father (H. O. Smith) had found the large clapper of a bell in the attic of a student rooming house they were demolishing in the 1950s on Beaver Avenue. Skip had saved the clapper and always wondered whether it came from the Old Main bell. As we spoke, half a century later, he believed it was probably still in a box in his garage. Would Gordon want it?

This astonished everyone but especially Gordon, because (as he told us) one of the vexing details in the Old Main bell project had been the absence of a clapper! I believe most people knew that the bell never rang, that the quarter-hour chimes were a recording, but apparently no one knew until the class project came along that it *couldn't* ring because, well, the clapper was in Skip's garage! Some research uncovered that it was once a regular student prank to mute the Old Main bell by swiping this essential part, and eventually the clapper did not return and was never replaced. Of course Gordon would like to have it! Skip found the clapper, and experts determined that it was, indeed, made for that bell.

In early May, Gordon, Patrick, Judy, and I gave a tour of the site to a number of folks from the administration. Afterward, most of us met in Old Main to discuss an anticipated shortfall in the arboretum's operating budget after opening, as Skip's gift was being paid off in installments. Rod Erickson and Gary Schultz agreed to backstop the budget with a $260,000 supplement in the next fiscal year.

In mid-June 2009, we hosted a late-afternoon preview reception at the Overlook Pavilion for everyone who had contributed to the arboretum project since the 1970s. Skip Smith was invited and attended. At least 150 people came, and we congratulated ourselves on our mutual success. Afterward, the staff and I hosted a core group for dinner at the Nittany Lion Inn, after which we all reconvened at my house for dessert and several hours of conversation and reminiscence. These were the individuals most involved in the still-ongoing building project: Gordon Turow, Dave Zehngut, Judy Larkin, Jeff Dice, Katie Rountree,[8] Missy Marshall, and Bob Vukich. It had been raining and chilly for several days, but somehow the weather cleared just as the reception began that afternoon and remained pleasant until the last of our guests departed in the evening.

Construction was mostly completed by early August. The construction fences were removed, and visitors gradually began to show up. Before classes began, we purchased several park benches and placed them around the gardens. We did not bolt them down, as is customary on campus, because we expected to move them as we learned where visitors preferred to sit. Police Services warned us that the benches would quickly disappear after the students returned, but they never did. Eventually, some were bolted in place simply because we did not want them moved, but neither theft nor vandalism has ever been a notable problem in the gardens.

We dedicated the H. O. Smith Botanic Gardens on Sunday morning, April 25, 2010, during the weekend of the public kickoff of the "For the Future" capital campaign.[9] It was a damp and unseasonably cold day. Although 858 people had preregistered, I was sure attendance would be light in such gloomy weather. However, the rain had stopped by the time of our scheduled start at 10:00, and over 600 people were present for the event. This was the first year of what became an annual display of twenty thousand to thirty thousand tulips and other plants around the Esplanade, and the flowers were in full bloom for the occasion, as planned. We had set up refreshment stations and displays to attract interest throughout the site, and many people lingered for a good while before heading home.

The project has received several awards. Overland Partners won a Texas Society of Architects award for their design of the Overlook Pavilion, one of 11 such awards out of 201 entries in 2010. That same year, at the State College Borough's annual ABC Dinner, we were awarded the Ingrid P. Holtzman Award for "substantially enhancing the aesthetic appearance of the Borough." M·T·R's design received the 2011 Honor Award for General Design from the Pennsylvania/Delaware Chapter of the American Society of Landscape Architects. In 2013, the project was recognized with a "Silver" LEED certification, and in 2020, the arboretum was one of three public spaces to receive the Pennsylvania Great Places award from the Pennsylvania Chapter of the American Planning Association.

The Arboretum Becomes Operational

Until 2008, the arboretum staff consisted of only Kate Reeder and me. However, when Phase I construction began, hiring additional staff became both necessary and possible. Skip Smith's gift pledge would soon generate an endowment income large enough to support salaries, and with the leadership gift in hand, we were free to fundraise for more projects within the arboretum.[1] A staff member dedicated to raising funds would be a boon for the arboretum, and Rod Kirsch fulfilled an earlier promise to provide us with a development director. All development positions were centrally funded and reported to the Office of University Development (in this case, to associate vice president Dave Lieb) while working hand-in-glove with the local administrator (in this case, the arboretum director) to raise funds for the unit. The position was advertised, and we hired Patrick Williams, who had been an associate director of development in Corporate and Foundation Relations.

Patrick started the job in January 2008 and occupied an office near mine in the Forest Resources Building. He faced the challenging task of raising money for a new program with no alumni and a very tiny record of prior donors. At the outset, his only resource was a list Kate and I had created of about twelve hundred people who had shown at least casual interest in the arboretum over the previous eight years. Patrick's personal affinity for the project and his sincere consideration for the needs and happiness of our donors made him perfect for the job. He remained our director of development through my retirement and beyond.

Rod soon agreed to fund 50 percent of a second position to assist Patrick if I could fund the balance from other sources. The other source became the Arboretum Steering Committee, whose members each supplemented the budget with contributions of about $5,000 annually. The supplemental funding became unnecessary after three years, as the endowment grew with Skip's pledge installments. In addition, new fundraising was going well as Patrick settled into his job. By the end of 2010, the endowment had grown by $1.6 million in additional gifts and pledges.

We hired Cathy Kennedy in April 2008 as a half-time development assistant and half-time budget assistant. She was a welcome addition to our small staff and quickly embraced the spirit of our new enterprise. She cheerfully adjusted her hours to work during evenings and weekends whenever the need arose, which it often did. Until her retirement in 2014, Cathy took the initiative to produce a two- or three-page summary of the week's highlights every Friday afternoon. This became a valuable historical record of the early years of the arboretum. As I write this, she has continued her involvement with the arboretum as a volunteer docent.

After Cathy's retirement and a series of interim appointments, her former responsibilities were divided between two positions created in 2016, when it became clear that we needed a dedicated budget assistant. We hired Yuhui "Millie" Zhou to fill the budget position, while Kelly Oleynik took on the roles of half-time development assistant and half-time volunteer coordinator. Kelly had been one of our first two Children's Garden interns in 2014 and continued until 2016 as a part-time development assistant. Cathy, Kelly, and Millie were all steadfast employees of the highest caliber, and we were blessed to have them as part of the arboretum team.

All of the plants ordered for Phase I construction were scheduled for installation in 2009, and it was essential to hire a staff horticulturist before this took place. Because the horticulturist's office was in the Overlook Pavilion after Phase I was completed (with the other staff members being housed on campus), this position became central to nearly all arboretum operations, including collections, seasonal displays, maintenance, volunteer programs, events, and ecological restoration work. However, in 2008, I was most in need of a person to monitor the quality of plant material as it was installed and begin assembling plant accession records from information available only temporarily from our contractors and on nursery labels. We advertised the position late in 2008 and hired Rick Johnson from a pool of nearly one hundred applicants.

As early as spring 2009, campus offices began contacting Kate about renting the Overlook Pavilion for events when construction was done. Later that year, we conducted our first event rentals at the Overlook Pavilion for the Penn State Alumni Association (staff picnic), Department of Architecture (staff luncheon), Board of Trustees (tour and buffet), Landscape Architecture faculty and graduate students (picnic), Schreyer Honors College Advisory Board (dinner), and the Alumni Association Executive Board (dinner).

Around the middle of 2008, Kate had suggested getting the Berkey Creamery to create a special ice cream flavor for the arboretum. Tom Palchak, the manager of the Creamery, said he would consider it, so we polled the faculty, staff, and graduate students in my academic department for flavor and name suggestions. From dozens of suggestions, we chose the entry by Emily Hill, a staff assistant, who suggested a vanilla-mint flavor with a berry swirl and the name "Arboretum Breeze." Tom added dark chocolate bits, changed "berry" to raspberry, and found a recipe that pleased his staff. The Creamery made its first large batch in September 2009, and

Tom set aside twenty tubs for the dedication of the H. O. Smith Botanic Gardens in April 2010. Arboretum Breeze remains one of the flavors in their standard rotation and seems to be in stock several times a year.

Not long after classes began in fall 2009, a group of students in the Schreyer Honors College approached us about organizing a music festival on the Event Lawn. They were working on a community service assignment in Judy Ozment's Leadership Jumpstart class for first-year students. We liked the idea. It would be the arboretum's first public event, and we were pleased that the arboretum was so quickly integrating into the academic life of the University.

The music festival, featuring local volunteer talent, took place on the afternoon of October 25—a perfect autumn day, as it turned out. A stage was erected on the Event Lawn, and free popcorn and ice cream were made available at the Overlook Pavilion. Hundreds of students showed up over the course of the afternoon. We invited Skip to come, and I remember being impressed by his unassuming demeanor as he patiently waited for a bag of popcorn. Then, unnoticed and seeking no notice, he slowly made his way through the crowd to the bottom of the Event Lawn and back. He was happy and did not seem to mind that not one of the students knew he was the person who had made the arboretum possible.

We worked hard in the first year to establish policies that were reasonable, fair, and consistent with the arboretum's mission[2] and University policies. We modeled simple rules of behavior for visitors after the practices of other arboreta and botanic gardens. Typical requests for venue rentals (weddings and departmental events at the Overlook Pavilion) were easy to fulfill, but we could not rent space for *any* purpose. After the first year, for example, we decided against hosting any more wedding receptions, which we naively thought *our* clientele would never carry to the excess of the proverbial Irish wake. We hosted public events organized by University units for free if they seemed appropriate for the venue and our mission and were not better suited to another campus location. Novel requests were discussed in staff meetings using the criteria of precedent, principle, and fairness.

With one or two minor exceptions, no one ever objected to our decisions not to rent or provide arboretum space for specific events. However, after many years, we received a request from the faculty adviser for a student group that wanted to hold a public event with clear political overtones. Usually, this would have received an easy "no" (inappropriate for the venue / better done elsewhere). However, given the possibility of an activist backlash, I consulted with the provost, Nick Jones. Nick brought the Office of General Counsel into the discussion, and my inclination to decline permission for the event was reaffirmed, but they took the occasion to assign an attorney to review our policy document.

She had many good suggestions, but I was disappointed in her dogged determination to reduce our event policies to simple, black-and-white rules. This was fair enough in principle, but it led to the removal of *all* discretion from our policies, which I thought was beyond what was wise in the larger interests of the University. I believe it is unhealthy to so rigidly regulate human matters that accountability and the exercise of judgment and conscience are always preempted. Most surprisingly to me, she struck all provisions for cosponsored events organized by students and student organizations, reasoning that they were non-University entities. Unless we *rented* our spaces (which the students could not have afforded), our revised policy would not have permitted the Schreyer students' music festival described above, or a student-organized candlelight vigil that we sanctioned on the occasion of a tragic accident. Perhaps there is an insurance risk associated with student-led events. But some risks are worth taking, and these could have been evaluated on a case-by-case basis by the Office of Risk Management, whose advice we sought on many occasions.

Fortunately, this was the only time our operations were seriously questioned. We worked hard to operate responsibly, conscientiously, and fairly. Kate was scrupulously attentive to University policy, and I depended upon her for that. The arboretum and the University benefited from the relative freedom from fussy oversight we enjoyed during those years, because "yes" can often move you forward when the easiest answer in a bureaucracy is an unconsidered "no." But I suspect we were viewed suspiciously as something of a "free radical" (a metaphor from chemistry), given the emerging prominence of the arboretum and its nebulous fit within the University's administrative structure, and the arrangement changed after I retired.

Our first wedding rental took place at the Overlook Pavilion on Saturday, May 8, 2010. Kate was there, as usual, to ensure everything went smoothly. As the time approached, I grew concerned that the wind was dangerously strong for our wedding party. It is often particularly windy at the Overlook Pavilion, and what I observed at home in Boalsburg seemed strong enough to cause problems with the heavy curtains around the event terrace. My wife, Susie, and I drove to the arboretum to help Kate in whatever way we could. Kate's husband, Tim, had already been summoned and was there when we arrived. The wedding was in progress as the wind gusted to over fifty miles per hour (according to weather records), and the curtains had torn loose from their anchors in the pavement. The four of us struggled outside the pavilion to hold the curtains down as the bride, groom, and rabbi stood inside and downwind just a few feet away. It was more than a little chaotic, but I suspect the couple and their families remember the wedding vividly and tell fond stories of their shared adventure! Such was our introduction to the marriage business. We got better.

Rick Johnson, the staff horticulturist, assisted eagerly as we completed Phase I and commenced our first full season of operations, but he remained with us for just one year. I sought a national pool of applicants when we began the search for his replacement in 2010. We particularly needed a horticulturist who had worked at another public garden, as none of us had that experience. To supplement the usual HR job announcement process, I solicited names of potential candidates from a couple of dozen colleagues around the country, and I encouraged applications from the people they recommended. This resulted in a strong pool of more than thirty applicants, and in due course we offered the position to Andrew Gapinski. Andrew had a baccalaureate in horticulture from the University of Wisconsin, a master's degree from the Longwood graduate program in public horticulture, employment experience at the Morton Arboretum, and volunteer experience at two other well-known public gardens.

Andrew put the operations of the arboretum on an excellent footing, not only because he was industrious and competent but also because his open, friendly manner made him universally popular. He established good relationships with our OPP landscape maintenance crew, whose quality of work was central to the arboretum's success. His rapport with OPP staff enabled us to supplement their contracted maintenance with the services of volunteer gardeners, whose labor contributions could have been a point of contention with the employee union. Unfortunately, Andrew was with us for only three years before the Arnold Arboretum hired him away. I was proud of him and even more pleased in 2023 when he became the executive director of the University of Minnesota Landscape Arboretum, one of the most prominent university arboreta in the country.

Our 2013 search for Andrew's replacement yielded an even larger number of applicants than the 2010 search. We hired Shari Edelson from this pool. Like Andrew, she had a master's degree from the Longwood program and work experience at several arboreta. Shari and Andrew were very different but equally competent and excellent additions to the team. During her nine years in the position, Shari did much to advance the arboretum's horticulture, displays, volunteer program, educational outreach, and public programs. I eventually changed her job title to director of operations. Sadly for us, she left in 2022 to pursue a PhD at Penn State in the interest of advancing her goal of someday becoming an arboretum director. Unless her interests change, I believe she will succeed.

We were eager to create programs and events that would establish the arboretum as a vital part of the University. This became easier as the staff grew, but some events still stretched our capacity. This was especially true of our Arbor Day programs and Pumpkin Festival, inaugurated in 2010 and 2011. The staff burden was eased in 2012 when we hired Alisha Barton as a part-time curatorial assistant. Alisha was a young Penn State graduate from nearby Unionville, where she distinguished herself as a member of her town council even as an undergraduate. She briefly left us to pursue other employment, and we hired her back full-time in 2015 to curate and map[3] our collections and coordinate large public events. She excelled in this work and became a key staff member. Nowadays she coordinates reunions of former staff members.

The Arbor Day event featured nature-related demonstrations and activities conducted by a volunteer corps of around forty experts drawn from the faculty, graduate school, and community. We conceived and organized the event in collaboration with Donnan Stoicovy, then the principal at Park Forest Elementary School in State College. On Donnan's suggestion, we targeted fourth-grade elementary classes and invited all of the State College elementary schools. I regretted that we could not extend the event to other school districts, but it was all we could do to accommodate between four hundred and six hundred fourth graders from State College alone. With about twenty stations and a similar number of fourth-grade classes, we resorted to an elaborate schedule of timed rotations and volunteer guides to ensure that each student benefited fully from the event. The residents' council of The Village at Penn State regularly gave us a large donation that covered the cost of meals for volunteers and the rental of tents, tables, and chairs.

Arbor Day in Pennsylvania falls on the last Friday in April, and the weather on that day is often chilly and damp. The children never seemed to mind, and their thank-you letters frequently mentioned the event as one of the highlights of their school year. Some teachers minded the weather, though, and last-minute cancellations marred the Arbor Day events in 2015 and 2016. Nearly half of the classrooms did not show up in 2016, leaving our presenters with gaps in their schedules. They were gracious, but we knew that volunteer participation would probably drop after spending half the day with nothing to do but try to stay warm. The staff was also dispirited after their enormous investment of effort. We could not think of a good way to adjust the schedule for no-shows or to discourage teachers from canceling their promised visits, so we decided to abandon Arbor Day as one of our regular events. An unstructured, less eventful production would have helped buffer against absences, but it would not have been as good, and that, too, would have discouraged attendance.

After our first year of arboretum operations, Andrew suggested hosting a festival modeled after a jack-o'-lantern carving contest his father regularly organized as a community service. The first Pumpkin Festival took place on a Friday and Saturday in October 2011 and has continued every year except 2020 (the year of the COVID-19 pandemic). The event quickly grew in size. After two or three years, we were giving away nearly a thousand pumpkins on the Sunday before the event, and most were returned as carved jack-o'-lanterns. The jack-o'-lanterns were lighted and displayed on straw bales around the Event Lawn, and a small team of campus and community dignitaries chose the winners by age category and special themes. We hired a band and invited a selection of commercial food vendors, and numerous activities and attractions were available for children. About ten thousand people typically turned out for the festivities. Private donations, supplemented by tourism grants, covered most of the annual cost of about $20,000. The first festival was sponsored by Dottie O'Connor, a former teacher and well-known patron of the arts. Mary and Tracy Vargo were the principal sponsors for about ten years while their sons attended Penn State. Most recently, Dennis and Sandy Hall, two long-standing arboretum volunteers and donors, have stepped forward as lead sponsors for the festival.

Other events during these years included spring plant sales (2012–15); the Winter Celebration, involving lighted displays and holiday songs on the last evening of fall classes (2014–19 and now resumed); outdoor art exhibitions organized jointly with the Palmer Museum of Art (2016 and 2018); Forest Fest, organized by Extension faculty (2011 and 2013); a solar eclipse viewing organized by the Department of Astronomy and Astrophysics (2017); and Music in the Gardens, organized by the School of Music as part of the Penn's Woods Music Festival (2013 onward, except for 2020). Most of these attracted a thousand or more visitors. Penn State departments organized many smaller public events and performances.

I have mentioned the private donations that enabled some of our events. The arboretum also benefited from gifts, often unsolicited, for other programs and activities. Several gifts endowed educational programs that were dear to the donors. Marcia Day supported the education of children, for instance; Donna Queeney, educational programs about birds; Sandy and Dennis Hall, programs in the Children's Garden; Mark Tygel, the horticultural education of youth; and Earle "Skip" Lenker, the enrichment of student internships. These endowments and other gifts directly sustaining educational programs helped establish the arboretum as an educational asset and facilitated its eventual transition to an administrative home within Penn State Outreach.

The Lenker endowment deserves special mention because it helped leverage our OPP maintenance contract for educational purposes. Because of seasonal fluctuations in workload, OPP needed to hire wage-payroll labor during the summer and autumn months to perform our maintenance. Instead of following the usual hiring process, Jeff Dice proposed converting the positions into competitive internships for undergraduates. Our horticulturists, Andrew Gapinski and later Shari Edelson, advertised the program to students, and we typically received around thirty applications for a handful of positions. A joint committee selected the interns, Jeff employed them, and our horticulturist coordinated weekly educational enrichment programs. The maintenance contract covered their wages, and the Lenker endowment funded special activities such as trips to Phipps Conservatory in Pittsburgh and the Scott Arboretum at Swarthmore College. The program has launched several students into public garden careers, including Alisha Barton, who began her employment with us as an OPP intern in the summer of 2011.

Another source of donated support came from the volunteers who added so greatly to our capacity. In 2019, for instance, nearly nine hundred volunteers contributed over seven thousand hours of assistance in staffing the Pumpkin Festival and Winter Celebration, guiding tour groups, supporting educational programs in the Children's Garden, assisting our OPP crew with garden maintenance, greeting visitors on football weekends, and participating in ecological restoration work. In round numbers, and on average, our volunteers enabled the arboretum to annually provide 120 drop-in programs for children, accommodate 50 public school visits, give tours to 100 adult groups, and staff 5 public events and festivals. We showed our gratitude to volunteers with special social events and bus tours to distant gardens, and we included them in more general (but invitation-only) events such as garden dedications. In 2017, we hired Rebecca Horwitt to assist Kelly Oleynik in organizing these events and developing materials and training to enhance the effectiveness of the volunteer program.

The arboretum employed 10.75 full-time equivalent staff when I retired in 2021, not including the maintenance personnel assigned full- or part-time to the arboretum through our contracts with OPP. The entire team was enthusiastic about the arboretum and always ready to meet the challenge of stretching our capacity when occasions demanded. Our ready commitment to public service led to some unforgettable encounters. After a donation enabled us to acquire the equipment, we began engraving plant labels that were then mounted on stakes and placed in the gardens. A student once came to us with a request to fabricate one of these with a brief message to his girlfriend, asking for her hand in marriage. The idea was for us to place the "label" in a certain flower bed to be "discovered" when the couple came for a stroll through the gardens. We obliged, much to the young woman's delight, and her answer was "yes!" How could we disappoint such a bold and enterprising young man?.

Whatever the reason—perhaps the satisfaction of working at a popular addition to campus—the arboretum staff was a motivated bunch of "happy warriors." Collectively (and of course that is how a team works), they displayed a rare and distinctive vitality. Working alongside these capable people with such passion for their work has been one of the most rewarding experiences of my life.

1. The Mitchell Tract as it appeared in February 2002 from an upper floor of the Agricultural Science and Industries Building. Student Parking Lot 80 (since replaced by the Business and Forest Resources Buildings) is visible in the foreground. Photo by Kim Steiner.

2. Penn State personnel and employees of Green Acres Nursery excavating the Hosler Oak in March 2005. Left to right: Chuck Biemesderfer (Green Acres Nursery), Drew Beben, Corby Fetterolf, Jeff Dice, unidentified, Ryan Greeley, Jim Sellmer, unidentified, and Kris Edson. Photo by Howard Nuernberger.

3. Planting the Hosler Oak on March 21, 2005. Pictured with shovels, from left to right, are Kim Steiner, Rod Kirsch (partially hidden), Gary Schultz, Rod Erickson, Eric Barron, Bob Steele, Charles Hosler, and George Biemesderfer (in helmet). Photo by Jim Sellmer.

4. The Hosler Oak under tender care in fall 2005, irrigated and monitored for signs of stress by Jim Sellmer and Dave Despot. Photo by Sam Grinstead.

5. The Mitchell Tract in July 2007, with distant portions of the arboretum in the middle distance. The Air Quality Learning and Demonstration Center is faintly visible near the center of the photograph at the edge of a cultivated field. The Hosler Oak witness tree is the single tree in a field of brown grass about halfway between the AQLDC and the Business Building. Photo by Kim Steiner.

6. Groundbreaking ceremony for Phase I on November 2, 2007. Left to right: Bob Steele, Kim Steiner, Graham Spanier, Skip Smith, and James Broadhurst. Photo by Greg Grieco.

7. The first tree planted during Phase I construction (March 31, 2009). Photo by Kim Steiner.

8. VIP tour of Phase I construction in May 2009. Clockwise from lower left: Barb Christ, Gordon Turow, Al Horvath, Judy Larkin, Steve Curley, Bob Steele, Gary Schultz, Rod Erickson, Kim Steiner, and two unidentified women. Photo by Rick Johnson.

9. Aerial view of Phase I construction in November 2008. Photo by Gregg Kohl, AC Photo.

10. The H. O. Smith Botanic Gardens as it appeared days before the dedication in April 2010. Photo by Andrew Colwell.

11. Aerial view of the H. O. Smith Botanic Gardens in early October 2018. Photo by Barton Malow Engineering.

12. Arboretum dedication ceremony on April 25, 2010. Left to right: unidentified student, Rod Erickson, Steve Garban, Missy Marshall, Kim Steiner, Graham Spanier, Skip Smith, Bruce McPheron, Bobbi Korner, and two unidentified men. Photo by Cody Goddard.

13. Kate Reeder and Andrew Gapinski at the "pumpkin giveaway" for the second Pumpkin Festival (2012). Kate and Andrew were among the first employees to join the arboretum staff. Photo by arboretum staff member.

14. Staff at the "pumpkin giveaway," prior to the 2018 festival, posing for a thank-you card to arboretum volunteers. Left to right: Kim Steiner, Alisha Barton, Rebecca Horwitt, Kate Reeder, Yuhui "Millie" Zhou, Patrick Williams, Shari Edelson, and Kelly Oleynik. Photo by Linda Duerr.

15. Jack-o'-lantern judges for the 2015 Pumpkin Festival. Standing, from left to right, are Damon Sims, Molly Barron, Steve Bair, Rick Roush, Bobbi Korner, and Donnan Stoicovy. Shari Edelson is kneeling in costume. Photo by Kate Reeder.

16. Cathy Kennedy, an early staff member, assisting at the 2011 Arbor Day event for local elementary students. Photo by arboretum staff member.

17. Students from twenty-four fourth-grade classrooms in the State College Area School District at the 2014 Arbor Day events. Photo by Kate Reeder.

18. Attendees at an annual volunteer appreciation event in 2016. Pictured staff include Kate Reeder (far left), Alisha Barton and Kim Steiner (standing at far right), Shari Edelson and Linda Duerr (seated on step at far right), and Patrick Williams (on seat wall at far right). Susie Steiner is the second person standing from left. Photo by Kelly Oleynik.

19. The creation of a xeric limestone prairie in the arboretum began in 2010 with the removal of brush on this hillside into Big Hollow, later followed by the planting of native prairie plants. This photo shows volunteers at work in April 2011. Photo by Andrew Gapinski.

20. The aftermath of a controlled burn on the prairie site in March 2020 (top) and regrowth the following summer. The burn was achieved in stages, managed by plowed fire lanes (visible at the right and top). Photo by Jonathan Chester.

21. The volunteer crew that conducted a controlled burn of the prairie site in March 2020. Jennifer Case (in red helmet) and Pat McElhenny (at far right), both with The Nature Conservancy, organized and supervised the burn. The rest of the crew were on loan from the Pennsylvania State Game Commission to accumulate experience for certifications. Photo by Shari Edelson.

22. Rod Erickson, Kim Steiner, Helen Hintz, Ed Hintz, Keith Masser, Skip Smith, and Rod Kirsch at the groundbreaking ceremony for the Childhood's Gate Children's Garden on May 30, 2013. Photo by Kate Reeder.

23. A “team-work minute” in October 2013 during construction of the Children’s Garden. Back row, left to right: Patrick Williams, Brian Phiel, Linda Duerr, Shari Edelson, Kate Reeder, Katie Rountree, Mark Saville, John Bechtel, Mike Arnold (AECOM), and Derek Kalp. Front row, left to right: Kim Steiner, Emmanuel Didier, and three employees of Leonard S. Fiore, Inc.: Sam Seltzer, Lloyd Foor, and Jeff Lucas. (Signatures were later hidden by additional stonework.) Photo by Judy Larkin.

24. Aerial view of the Children’s Garden in October 2018. Photo by Barton Malow Engineering.

25. Skip Smith and sons John (left) and Dave (right) at the Children's Garden dedication ceremony on September 7, 2014. Photo by Kate Reeder.

26. Setting the "digit" stones during construction of the Joel N. Myers Sundial in July 2011. Mark Mennin, the sculptor, is standing at left. Photo by Andrew Gapinski.

27. The Joan Milius Smith Esplanade and Joel N. Myers Sundial in 2017. Photo by Steve Tressler.

28. A view of the H. O. Smith Botanic Gardens from the top of the Business Building on the occasion of the 2014 Music in the Gardens event organized by the School of Music. Photo by Cody Goddard.

29. An evening view of the Overlook Pavilion. Photo by Cody Goddard.

30. The Overlook Pavilion's watershed map (Ridge and Valley Sculpture). Photo by Fred Weber.

31. The Hosler Oak witness tree in the autumn of 2010. Photo by Andrew Gapinski.

32. The Margery Enes Smith Fountain is a popular backdrop for spring graduation photos. Photo by Steve Tressler.

33. Display gardens on the perimeter of the Esplanade in August 2012. Displays were changed seasonally, beginning with tulips and often other bulb plants timed to be at their peak of flowering during spring graduation. The gourd arch, apple trellis, and flower beds in the center were removed in 2019 to create the main entrance into the Pollinator and Bird Garden. Photo by Andrew Gapinski.

34. A view toward the Esplanade from above the Round Meadow in the Pollinator and Bird Garden, October 2021. The gourd arch had been located near the center of the photograph. Photo by Rob Cardillo.

35. Lotus Pool on the Oasis Garden Terrace. The space for the pool was originally intended for a planting, but the name of the garden was retained even after the change in design. Photo by Steve Tressler.

36. A bower supporting wisteria vines in the center of the Rose and Fragrance Garden. Photo by Andy Stephenson.

37. View over the Marsh Meadow toward the wall enclosing the Rose and Fragrance Garden. Photo by Jeff Remington.

38. North Terrace Garden (also known as the Conservatory Terrace Garden) with the Bamboo Allée in the background. The Poplar Court, occupying the anticipated location for a conservatory, is to the left and behind the camera. Photo by Kate Reeder.

39. Children playing in the In and Out Creek of the Children's Garden. Photo by Kim Steiner.

40. Entrance to the Limestone Cave in the Children's Garden. Photo by Steve Tressler.

41. Staff members Linda Duerr (standing right) and Jenn Hooven (kneeling) with visitors in the Children's Garden. Photo by Kim Steiner.

42. Linda Duerr with children at the entrance to the Children's Garden. Photo by Kim Steiner.

43. The Pollinator and Bird Garden in the autumn following completion in 2021. Note the dead Table Mountain pine "raptor perches." Photo by Rob Cardillo.

44. Student watching the activity within an observation beehive in the Pollinator and Bird Garden. Photo by Rob Cardillo.

45. Pond, "birdhouse," and the observation beehive enclosure of the Pollinator and Bird Garden shortly after its completion. Photo by Rob Cardillo.

46. Research and demonstration plots within the Pollinator and Bird Garden. Photo by Andy Stephenson.

47. View within the Pollinator and Bird Garden. The structure at upper left provides nesting sites for various species of solitary bees. Photo by Rob Cardillo.

48. Arboretum staff gathering in December 2021, after the change in leadership and administration. Standing, left to right: Kelly Oleynik, Yuhui “Millie” Zhou, Kate Reeder, Patrick Williams, Rebecca Horwitt, and Mandy DiBella. On the bench: Kim Steiner, Tracey Huston, and Shari Edelson. Kneeling and seated: Heather Frantz, Tim Gould, Sandy Smith, and Jenn Hooven. Photo by arboretum staff member.

Incremental Enhancements

A Class Gift—The Marsh Meadow Boardwalk

Completing Phase I in 2009 enabled us to pursue other parts of the Mitchell Tract master plan as enhancements that could be funded and constructed more or less independently, and some were relatively inexpensive. All presented expansion opportunities for the arboretum if we could raise the money.

Class gifts were a potential funding source for small projects with appeal to the generosity of graduating seniors. The gifts were chosen by student ballot after a preliminary screening of proposals by a committee of students, staff, and faculty. The money came from graduating seniors who elected to pledge the remaining balance of a security deposit made at the time of matriculation. In fall 2009, Kate Reeder and I submitted a proposal for the Class of 2010 to fund a boardwalk over the Marsh Meadow feature.[1] We were optimistic, but it was a pleasant surprise to learn that our project was the overwhelming favorite among three proposals voted on by nearly three thousand seniors. We were even more surprised when the final tally of gifts ($197,693) was the most money ever raised for a class gift project. We had heard many compliments about the new arboretum, but the enthusiastic generosity of the Class of 2010 was a vivid sign of its popularity with the student body.[2]

We contracted with the Bohlin Cywinski Jackson agency in Wilkes-Barre to design the boardwalk. Completion was delayed for some months when the University administration suspended action on uncontracted projects (owing indirectly to a budget battle in Harrisburg). However, construction finally began in August 2012, and the boardwalk was dedicated in a small ceremony on June 22, 2013.

Patrick Williams and I learned in July 2009 that Joel Myers, a Penn State alumnus and the founder of AccuWeather, had proposed to the campus Public Art Committee that he donate a sundial to the arboretum. This information came from Joyce Robinson, a member of the Public Art Committee. The committee liked the idea, and I did as well.

We estimated the construction cost and settled with Joel on the size of his gift. Our goal was a landscape-scale installation with an unconventional sculptural form that would function as a sundial with reasonably high accuracy. To ensure compliance with the standards of the Public Art Committee, I asked Jan Muhlert, director of the Palmer Museum of Art, to serve on a project oversight team. Dave Zehngut investigated potential designers through an internet search, and after soliciting proposals and conducting interviews, we selected Mark Mennin for the contract. Mark's specialty was "soft" sculptural forms in heavy pieces of granite.

Part of the reason we chose Mark was that he had already designed and executed two sculptural sundials in Connecticut. However, he did not pretend to be a sundial designer, and it was understood that the geometric features of the design would be our responsibility. Chris Palma (faculty, Department of Astronomy and Astrophysics) volunteered to provide that expertise. After Chris gave me a crash course on sundial design, we agreed upon stone "digits" marking the tip of the gnomon's shadow at selected hours in three rows: one for the summer solstice, one for the equinoxes, and one for the winter solstice. We also wanted to represent the analemma in a rudimentary fashion.[3] (A complete representation of the analemma would be possible on a plaza, but our sundial would be embedded within a lawn.) In addition to marking time on daily and seasonal scales, as sundials always do, we decided to give a nod to the geologic time scale with a separate stone placed in the direction of true north as it was sixty-five million years ago. Because our continent has moved and rotated since then, the direction in which the gnomon points (north by design) is nowhere near the direction of this stone. These were the cardinal elements of the design for timekeeping purposes; the rest was up to Mark.

Chris made the precise calculations for our position on Earth, and Rob Rath of Sweetland Engineering translated the data into survey points for construction. Mark executed all this in stone in his unique fashion.[4] Plans called for the top of the gnomon to be over eight feet high. Since this represented a potential climbing hazard, we consulted the University's risk officer, Gary Langsdale. Mark made some changes at Langsdale's behest, and none of us on the staff ever saw someone successfully climb the gnomon—or even try. However, about a year after the sundial was finished, Rod Erickson, then serving as president of the University, innocently emailed me a photo of his granddaughter sitting near the top of the gnomon. So it seems it *can* be climbed, and Rod's granddaughter has probably not been the last to do it.

The limitations of our sculptural approach mean that the inherent accuracy of the sundial can be precisely checked only on specific dates and on the hour. I have checked several times, and the sundial was always within a minute or two of the correct time. However, clock time must usually be roughly interpolated, causing precision to drop accordingly. It is always necessary to add eleven and a half minutes because we are west of the center of our time zone (that is, it takes solar time longer to catch up with clock time). We could have shifted the stones to cancel that deviation, but it seemed more appropriate for an arboretum sundial to align with the heavens rather than with time zones that did not exist before 1883.

We dedicated the sundial on September 22, 2011, with formal remarks before a crowd of nearly two hundred.[5] For a long time afterward, Chris and I were the only two people in the world who could explain the sundial, but we eventually drafted a description that was made into a sign. The workings of the sundial are not easy to grasp, but a mere lawn ornament with a simpler design would have been intrinsically inaccurate and would have foreclosed an opportunity to educate and learn.

My only regret about the sundial is that we placed it on the Joan Milius Smith Esplanade. The lawn of the Esplanade seemed like an ideal location, and the Arboretum Design Committee concurred. Unfortunately, it did not occur to me to consult Skip Smith. Joan Milius Smith was Skip's late wife, and he regarded the sundial as an intrusion rather than an enhancement to the Esplanade. Regardless of whether I should have anticipated this reaction, I certainly erred in failing to consult him in advance. We eventually got past that, but I still regret the clumsy and needless affront.

Restoration of Natural Areas

Even as our attention was on the 55 acres covered by the Mitchell Tract master plan, opportunities to improve the remaining 315 acres were never far from our minds. Phase I construction took place on the front half of the Mitchell Tract. The back half is an open field that slopes down to Hartley Wood and Big Hollow. Beyond the field, the arboretum's most significant developed features were the Bellefonte Central Rail Trail, the Air Quality Learning and Demonstration Center, and the American chestnut breeding orchard. However, none of these was likely to entice a large number of visitors to cross the open field between the gardens and Big Hollow. Almost from the beginning of the arboretum project, we embarked upon a couple of ecological restoration projects using volunteers and our meager resources to develop the natural features in remote parts of the arboretum.

Of the arboretum area not addressed by the Mitchell Tract master plan, approximately 245 acres are programmed in the Sasaki plan as "natural areas." As described earlier, much of this land is actively pastured or covered with regrown brush following pasture abandonment. About 60 acres consist of forest tracts formerly managed by the forestry faculty under plans developed in the 1920s and 1930s (appendix B). These tracts are located in small parcels along Big Hollow Road and a larger area directly between Big Hollow Road and Interstate 99. About half of the wooded acreage is covered with natural stands of oaks and other hardwoods (regrown after cutting by former owners), while the rest began as old fields planted around 1928 with various conifers. These are significant as natural areas, but even light management of them has been beyond the capacity of the arboretum staff.

Another 30 acres make up Hartley Wood, part of the 1989 Mitchell Tract purchase, located just downhill from the Overlook Pavilion.[6] Hartley Wood is of special significance because of its appeal as the closest woodland to campus and because it possesses characteristics of an old-growth forest.[7] At least seven trees in the woodlot are known to predate European settlement, and the age of one shows that it germinated in the 1650s. Trees that old are rare in central Pennsylvania because of its history of logging, charcoaling for iron furnaces, and farm woodlot utilization. We believe that some trees were harvested from the woodlot in the nineteenth century, but it was never completely logged. Setting aside rare, prehistoric occurrences of destruction by fires or tornadoes, Hartley Wood has existed as a mature forest since the last ice age.

Although Hartley Wood is interesting as a remnant of presettlement forests, its present ecological condition is degraded by a lack of the young trees needed to sustain its ancestral composition. Ground cover vegetation is dominated by non-native shrubs and other plants that have invaded from surrounding residential properties and campus. Non-native tree species are abundant as seedlings and are encroaching into the mid-story and lower forest canopy as trees. Native wildflowers that should be common are largely absent because of the shade cast by plant species that were not here two hundred years ago. The woodlot is fairly choked with underbrush, an unnatural condition attributable to the cumulative effects of European settlement.

Restoring Hartley Wood to something resembling its primeval state would require more resources than the arboretum staff can muster, but we have made progress in that direction. On a Saturday in March 2001, several of us (Neil Korostoff, Ken Tamminga, some Landscape Architecture students, and I) made a feeble but earnest attempt to remove the invasive shrubs from a small portion of the woodlot. It was more of a trial work session than an actual start on restoration. Even if we had persisted for the remainder of the semester, which we did not, the scale of our efforts would have been grossly inadequate for the task.

In 2006, Kate and I organized a volunteer program called the Arboretum Woodland Restoration Corps (AWRC) to continue the removal of invasive shrubs. A graduate student, Sarah Melissa Witiak, initially led the program with the help of Paul Lupo, a research assistant in the School of Forest Resources. The work was guided by a restoration plan developed by Sam Grinstead.[8] Chip Clark, a local citizen, succeeded Sarah as leader of the AWRC when she left in 2008 to take a job in Virginia. Sarah and Chip held work sessions every month or two in the fall and spring, with eight to ten or more volunteers attending each session.

Although Chip continued to volunteer for at least ten more years, Andrew Gapinski took over the coordination of the group in 2010 when he joined the staff as our horticulturist. We changed the name to the Arboretum Ecological Restoration Group (AERG) as the focus shifted to include an area adjacent to the woods where we hoped to re-create a prairie (discussed below). Over time, the AWRC and AERG improved trails and removed large amounts of unwanted brush, which was gathered into enormous piles in an open area at the bottom of Big Hollow and burned with the assistance of Alpha Fire Company.

In 2014, we received a substantial financial gift from Henry Gerhold, a retired faculty colleague, to reestablish native wildflowers along the trail system in Hartley Wood. Henry's gift, given in memory of his late wife, provided over ten years of support to improve the trails (appendix J) by removing invasive, non-native underbrush and planting tens of thousands of native wildflower plants. This work was organized and planned by Shari Edelson and executed by a succession of part-time project assistants whom we hired with Henry's funding. We also used student volunteers;

several dozen often turned out at a time for weekend work sessions.

The last work plan I reviewed had projected the reclamation of about twelve acres through 2025 (about 40 percent of the portion of the woodlot owned by Penn State). Whether the gains can be held against persistent invasion remains to be seen. Much of the character of our presettlement forests was shaped by occasional ground fires. Complete success in the battle to achieve and retain the natural character of Hartley Wood vegetation may not be possible without controlled burns.

This indirectly introduces the second of our ecological projects: re-creating a xeric limestone prairie. In our area, prairies do not develop and persist without occasional fires and, depending on fire frequency, the presence of shallow, drought-prone soils. The impetus for this project came from the master's degree research of our first arboretum-sponsored graduate student, Daniel Laughlin, under the supervision of Chris Uhl in the Department of Biology.[9] Daniel documented the occurrence and composition of ten limestone prairie remnants in the Ridge and Valley physiographic province of Pennsylvania. Two were found nearby, one on University land in Big Hollow (south of the airport) and one in Penns Valley. Although all of the extant remnants are very small, totaling less than 2.5 acres combined, historical documents indicate a much more extensive occurrence of prairies at the time of European settlement. The Penns Valley remnant was once part of a prairie that is known to have extended for several miles between Spring Mills and Centre Hall.

I had long been aware of the former existence of prairies in Pennsylvania, but my interest in re-creating a prairie in the arboretum[10] was piqued when Daniel's research revealed that nearly all existing prairie remnants have a particular soil type that happens to occur in a small part of the arboretum. This five-acre site is a sparsely wooded hillside immediately east of Hartley Wood. It is just outside the Mitchell Tract, near the master plan's large Meadow Garden (appendix E).[11] A small prairie in this location would complement an axial "grass" theme that Missy Marshall began developing in 2002 and eventually (in the 2008 revision) extended from the Marsh Meadow along Park Avenue to the Meadow Garden by way of the Bamboo Allée, Event Lawn, and River of Grasses (downslope from the Overlook Pavilion).

To create a prairie on this site, we began in 2010 by clearing a nearly impenetrable thicket of privet, honeysuckle, and other shrubs that had invaded after the site was abandoned as a pasture some decades earlier. As work progressed, we benefited from the advice and assistance of many people. Chief among these were Daniel Laughlin, Harold "Hal" Gardner (a PhD alumnus of Penn State and author of the book *Tallgrass Prairie Restoration in the Midwestern and Eastern United States*), Jennifer Case and Patrick McElhenny (both with The Nature Conservancy's fire program), Harry Henderson (an amateur botanist with extensive knowledge of the local flora), Joel McNeal (a doctoral student in the Department of Biology), and Mark Fiely (a horticulturist with Ernst Conservation Seeds).

Creating even a small prairie is not a trivial project under any circumstances, and it would have been virtually impossible before 2009 because of the necessary use of fire in our humid climate. In that year, a new Pennsylvania law minimized the risk of penalty for damages or injuries from controlled burns provided that the burns are conducted and supervised according to regulations.

Prior to 2009, corporate and public landowners in Pennsylvania did not use controlled burns to manage vegetation; since then, the practice has become moderately common. Still, our location near a residential area and University buildings presented a special challenge, not so much because of risk but rather as a public relations issue. We were encouraged by the precedents set by the Nichols Arboretum (University of Michigan) and the University of Wisconsin Arboretum, both of which have conducted prescribed burns within city limits for many years.

Preparations for our prairie burns were elaborate and included creating firebreaks with student crews, working around weather conditions and fuel moisture levels, providing required notifications to agencies and local officials, and informing and educating the community. The Nature Conservancy's Jennifer Case supervised our burns as a credentialed "burn boss" and the Pennsylvania Game Commission provided crews, both at no cost to the University. The first burn was conducted in May 2013, and two days later we spread a prairie plant seed mix donated by Ernst Conservation Seeds. We burned the site again in May 2015, April 2018, and March 2020. The prairie was expanded to twelve acres in 2017 by seeding seven adjacent acres. Despite advance notice and publicity, the first burn caused alarm when smoke drifted across campus, and we were asked to extinguish the fire. We have had no further complaints, and prairie vegetation is gaining a foothold at the site as the burns have become almost routine.

A Home for the Staff

Designs for an Education Center, Conservatory, and Planetarium

As Phase I was under construction, Patrick Williams and I turned our thoughts to fundraising for unfinished parts of the Mitchell Tract master plan. Two that we focused on intently were a children's garden and a pollinators' garden (discussed in later sections). Here, I want to address the most expensive pieces that remained undone, the education center and conservatory, and new plans for a cultural district that followed from these.

Although it had not been in the master plan before, we were also considering a planetarium. Putting a planetarium in the arboretum had been suggested by Dan Larson (dean, Eberly College of Science) in an Arboretum Steering Committee meeting in 2004, and the idea came under serious discussion in October 2007 on the initiative of Larry Ramsey, who was head of the Department of Astronomy and Astrophysics. A planetarium is a natural fit for an arboretum, and there are numerous examples of this being done elsewhere. What made the idea particularly appealing was that the digital equipment in modern planetariums, in addition to projecting stars and planets onto a white dome, can be used to create immersive experiences for instruction or artistic performance in many academic disciplines. Ramsey and Larson believed they could raise the money to build a planetarium through their college's development office.

In January 2008, Rod Erickson gave us the green light to develop a preliminary program and cost estimate for integrating a planetarium into the arboretum project. This was carried out by an internal committee of seven, chaired by Deborah Blythe (later Howard), director of Facilities Resources and Planning. Blythe presented the report in June in a meeting with Rod Erickson,

Gary Schultz, Bob Steele, and Eva Pell. It was approved, and a planetarium became part of Missy Marshall's final version of the Mitchell Tract master plan in September of that year (appendix I).

Getting the funding to build a conservatory or planetarium was appealing, but my primary interest at the time was a building for arboretum offices, storage, classrooms, and events. The Phase I two-person office and small storage room in the Overlook Pavilion were woefully inadequate for the staff, supplies, and equipment needed to run an arboretum. Instead, most of the arboretum staff occupied offices in the Forest Resources Building as guests of my academic department, and we stored supplies in my research lab. Eventually, this grew to seven offices and additional storage space in my academic building and remained that way until 2023. The space to house our operations was a persistent problem, and the solution before us was the master plan's education center.

To raise money for the education center, we needed an updated cost estimate and a rendering that showed what a building with that price tag might look like. I broached the idea of funding a planning project with Dave Lieb early in 2011. Rod Kirsch, Ford Stryker, and Larry Ramsey were brought into the discussion, and it was decided in April to include all three of the expensive buildings (education center, conservatory, and planetarium) in a request for proposals. A planning oversight team was created.[1] Other than a commitment of $30,000 from the Department of Astronomy and Astrophysics, a funding source for the study remained uncertain. The cost was expected to be about $100,000.

Five firms submitted proposals by the end-of-June deadline, and they all sought a substantially higher fee than we had anticipated. We selected Gund Partnership to do the work and hoped to come to terms by negotiating changes in the project's scope. Unfortunately, the contract was still unsettled in early fall, and no one in the administration had come forward with funding to break the logjam.

In October 2011, I met to discuss the project with Graham Spanier, Rod Erickson, Rod Kirsch, and Gary Schultz. (Gary had retired and then returned as interim senior vice president for finance and business following Al Horvath's departure in July.) Graham was skeptical about the planetarium and suggested it be dropped from the project. No funding was promised, and the only resolution was that Rod Kirsch should discuss the project with his University Development Council.

My next involvement in administrative discussions about this project was a January 2012 presentation with Laura Cabo from Gund to the Board of Trustees Subcommittee on Architect/Engineer Selection. Lest the subcommittee pick and choose among the three buildings with an uncertain outcome, I emphasized that the greatest need was an education center that could house arboretum operations. After some discussion, the chair of the subcommittee, Linda Strumpf, suggested that the decision at hand was whether to proceed with preliminary designs for two building projects, an education center and a conservatory. Including a planetarium as a third building would be a secondary decision. The subcommittee recommended proceeding with all three.

I have no recollection or record of who committed the funding for the study. It was probably Rod Erickson, who had been appointed president of the University following Graham's resignation in November. The balance may have tipped during a December meeting of the Project Decision Review

Board, which is an advisory council to the president. In any event, the contract with Gund was finalized shortly after the board meeting in January 2012.

The Gund folks (primarily Laura Cabo and Eric Svahn) visited campus for workshops on five occasions in 2012. Several of us accompanied Cabo on a June field trip to the Cleveland Botanical Garden and Museum of Natural History to tour their conservatory and planetarium operations. The Gund report was completed in January 2013 after a December presentation to Rod Erickson, Dave Lieb, Rob Pangborn (interim provost), Paul Wangsness, and Ford Stryker.

Gund's main task was to thoroughly analyze programming, space needs, and costs for the three buildings. However, the most striking aspect of the report was Gund's vision of their style, arrangement, and location (appendix K). To my eye, these were beautifully compatible with the H. O. Smith Botanic Gardens aesthetic. The projected fundraising targets for design and construction costs, along with endowment supplements for maintenance and staffing, were $29.6 million for the education center, $19.3 million for the conservatory, and $12.5 million for the planetarium. These numbers were presented to the administration in the December 2012 meeting. Engineering and design considerations dictated that nothing could be built without the education center. This meant a nearly $30 million threshold to construct any part of the plan, and Dave Lieb was doubtful that we could get there with philanthropy alone.

Instead, a "Cultural District"

As Dave predicted, we never found a donor for the marvelous vision of the Gund plan, and it became merely the first leg of a winding path toward a home for arboretum operations. The eventual outcome was the construction of the new Palmer Museum of Art, which houses the arboretum staff and its operational needs on the first floor of an administrative wing. This has been a net benefit to the arboretum, the University, and certainly the Palmer. However, the building has disappointed many who love the arboretum, which warrants an explanation of how this came about.

In April 2015, I was summoned to meet with President Eric Barron, Nick Jones, David Gray (the new senior vice president for finance and business), Ford Stryker, Gordon Turow, and Barbara "Bobbi" Korner (dean, College of Arts and Architecture). The stated purpose was to discuss "A&A and the arboretum." This was not very descriptive, so I called Bobbi to find out whether she knew more. She did, and she kindly explained. The scheduled meeting had grown out of a discussion some weeks earlier among Eric, Bobbi, Rod Kirsch, and John Driscoll, an alumnus and prospective donor of a significant art collection. The existing art museum was running out of room for new collections, and Mr. Driscoll suggested a new building clustered with other cultural assets in a location downtown or perhaps near the arboretum. President Barron picked up on this and proposed creating a cultural district for the University and community within the arboretum, capitalizing on its popularity, accessibility, and proximity to Schreyer House.

The purpose of the upcoming meeting was to seek my acquiescence; all of the other attendees were already aware of the plan and "on board." Bobbi was almost apologetic as she explained all this, but she liked the idea of relocating the art museum to the arboretum. In addition to needing more space, the museum had a problem with accessibility, which was challenging for off-campus visitors because of its central location on campus. At that time, the art museum was logging about 35,000 visitors per year, including students, while visitors to the H. O. Smith Botanic Gardens numbered about 150,000.[2] She thought that both facilities could benefit from crossover attendance.

I believed the arboretum would continue to be a more popular destination than the art museum, in which case any incremental crossover effect would benefit them more than us. However, perhaps unreasonably, I still harbored lingering doubts about the arboretum's security from future neglect, so anything that would strengthen its integration and importance to the University had automatic appeal. Aside from this, we had always cultivated a close relationship with the College of Arts and Architecture, one of two University Park colleges (along with Agricultural Sciences) with which we had successful collaborations across more than one academic department. In addition to the abovementioned performances and exhibitions with the School of Music and the Palmer, we annually hosted public performances of Shakespeare by students in the School of Theatre, and we frequently worked with the Landscape Architecture and Visual Arts faculties to arrange temporary displays of student projects. One of our first projects was participation in a successful grant proposal by the Center for Performing Arts as the venue for a production entitled *The Secret Life of Public Spaces*. Having the art museum nearby could only strengthen these relationships, and the adjacency of other museums might lead to new collaborations.

Although the composition of a cultural district was somewhat vague when Bobbi and I spoke, I concluded that this presented a strategic opportunity for the arboretum.[3] The risk was that the arboretum would be overrun by a miniature version of the Oklahoma Land Rush as all of the campus museums and cultural facilities sought to relocate there. However, given my experience of OPP's exemplary planning process, this seemed manageable. I was amenable but prepared to oppose any attempt to "subdivide" the arboretum.[4]

The president opened our meeting on April 29 by explaining his vision and distributing a one-page summary. He outlined several predicates for his proposal of a cultural district: (1) inadequate space in the Palmer, (2) limited parking on campus, (3) the campus's large number of small, dispersed museums that get little visitation, and (4) the aesthetic issue of service-oriented buildings bordering the arboretum, which could be mitigated by developing the cultural district. He proposed creating an arts and museum district that "borders and is a part of the arboretum, hides the service buildings . . . and puts museums and theatre facing the [arboretum]."

Although I was receptive to the idea, I later regretted not asking Eric for a clear commitment to ensure that the arboretum would be an equal partner in planning decisions. I believe he would have readily agreed, especially with a supportive comment from Bobbi, which I'm sure she would have volunteered. In fact, the arboretum *was* a full partner in planning the cultural district over the next two years, when Bobbi was still dean and the key personnel in OPP (the assistant vice president for physical plant, the director of Campus Planning and Design, the university architect, and the project manager) were people who had worked with the arboretum for years and manifestly cared for its interests. Sadly, *every one* of those individuals had retired or left by late 2018, and the interests of the arboretum became secondary in the art museum project.

In June 2015, President Barron appointed a Cultural District Plan Steering Committee to "explore the feasibility of modifying the arboretum's master plan to create a new cultural destination at Penn State."[5] The charge letter included the language from the April meeting about the district being "a part of the arboretum." The aims were to identify the right programs for a cultural district, determine an optimal arrangement of program blocks, and establish their locations on the landscape. In addition to the arboretum, the possible components of a cultural district were identified as a conservatory, a planetarium, an art museum (replacing the existing Palmer building), a science-technology-engineering-mathematics (STEM) museum (consolidating and perhaps expanding existing collections), a theatre performance space (replacing one or more campus theatres), a restaurant, and visitor parking. A request for proposals was issued, and we selected the firm of Ayers Saint Gross (ASG) after interviews early in September.

Discussions with ASG about the project approach and scope of work had continued for several weeks without a contract when we learned, on October 29, that the University was withdrawing funding for the project because of a budget emergency. When we met that day, a decision had already been made within OPP to abandon the ASG contract and use the staff of Facility Resources and Planning (Deb Howard's unit) and Campus Planning and Design (Gordon Turow's unit) to prepare a feasibility document for review by the Cultural District Steering Committee. Rachel Prinkey, the OPP project manager, came to the meeting prepared with a detailed list of the tasks ahead and a plan for moving forward.

By this time, we had already been holding regular meetings to prepare for the consultant's work, and we had gathered preliminary information from members of the campus museum consortium. The committee members knew we could do much of the work "in-house" as well as ASG could, and it was resolved that the rest would just have to be our best effort. However, I was troubled about what "the rest" might look like. We had already learned that most members of the museum consortium were eager to move their collections to the arboretum, and most would probably present a case for more space than they already occupied. So I was especially worried that any plan developed by University staff, without a third party to deflect blame for a museum's not getting what it wanted, would be more expansive than necessary or desirable[6] and would needlessly occupy arboretum space.

Patrick and I took this concern to Skip Smith in a face-to-face meeting the day after our steering committee meeting. I asked him to fund the ASG study at a cost of up to $440,000, the high end for the contract still under negotiation. Skip agreed, but he pointed out that the gift would diminish any future donations he might make to the arboretum (as it probably did). Fortunately, we needed only $256,000 after receiving a revised proposal from ASG in November. In this way, ironically, the arboretum became the sole funding source for the cultural district study.

With the ASG study back on track, we finalized a questionnaire and sent it to sixteen STEM and art museum directors and six other museum consortium members with substantial collections but no designated museums. It was a comprehensive survey calculated to yield a great deal of data. The arboretum's response alone ran to nine pages. This provided much of the raw material needed by ASG to digest and assimilate into a plan. As Karla Aghajanian, ASG's lead planner, put it, their challenge was to figure out the right size and mix of programs that would "responsibly" fit within and enhance the arboretum to draw more visitors.

ASG proposed a complex comprising 230,000 gross square feet of building space. The art museum was projected to occupy about half of the programmable space (excluding shared areas), a 60 percent increase over its current facilities. The proposed space for STEM museums and collections also represented increases for those most likely to be relocated: the Earth and Mineral Sciences Museum, Matson Museum of Anthropology, Frost Entomological Museum, Ecosystem Science and Management natural history collections, and PAC Herbarium. Space needs for the arboretum education center, conservatory, and planetarium were incorporated from the Gund plan without change. A café, museum shop, and outdoor amphitheater were included, but a restaurant and an indoor theatre were considered impractical. The complete technical report was over eleven hundred pages long.

The ASG project was essentially completed with an August 2016 presentation to President Barron. As projected by ASG, the cultural district was a massive construction project with a high price tag ($282 million), and it would undoubtedly need to be funded and built in stages. To account for uncertainty, ASG proposed a phasing plan (appendix L) that allowed each stage of incremental build-out to stand alone as if nothing more were ever intended.

In fall 2017, the University hired Grenzebach Glier and Associates, a fundraising consulting firm, to assess the feasibility of raising funds for the cultural district. They conducted a series of tele-briefings by President Barron for selected participants, most of whom were identified as donors or potential donors. I attended the briefing held on December 8. During this session, Eric shared his vision of building upon the success of the arboretum to create a University cultural destination next to campus. He retold the story of our "lonely" witness tree as a precedent for an inspirational first step in the cultural district plan. He also mentioned that $30 million was already in hand from a Big Ten media contract announced in February. Following the presentation, Eric answered questions from the audience about the business model, time frame, benefits to educational programs, and so on. Many participants supported the plan, which they thought would easily attract donors and help recruit students and faculty. I was gratified to hear numerous expressions of concern about the plan's impact on the arboretum, to which Eric responded with assurances.

The Palmer Museum of Art

In early January 2018, I met with Erin Coe (the new director of the Palmer), Katie Rountree (now the facilities manager for Arts and Architecture), Steve Watson (the new director of Campus Planning and Design, now renamed Planning, Design, and Properties), and Rachel Prinkey (the project manager for cultural district matters). Rachel requested the meeting to review our space needs in preparation for a presentation to the president.

Rachel explained that Eric was advancing the art museum, including space for arboretum administration and operations, as a reduced-budget first phase of a future cultural district. This was a surprise, because an art museum was a third-phase building in ASG's plan.

I questioned this decision in an email to the president, and he responded that the art museum was being advanced because it had the most donor appeal. This was surely known even without a consultant's assessment, and Eric had been briefed twice on ASG's plan for the cultural district before it was finalized. The plan's studied arrangement of buildings, according to size and function, would have been very different if ASG and the steering committee had been directed to prioritize construction of an art museum. Placing a stand-alone art museum in the location proposed by ASG (down the hill toward Hartley Wood, alongside the Housing and Food Services building) would have made no sense if the art museum were always intended to come first. Moving the location of the largest building now rendered the rest of the Gund plan obsolete.

All was then quiet for six months as the president looked for enough money before committing to the project. Then, in late June 2018, Bill Sitzabee (the new associate vice president for facilities management and planning, replacing the retired Ford Stryker) called a meeting of Rachel, Erin, Katie, Steve Watson, Lisa Berkey (Rachel's supervisor in Design and Construction), and me to announce the president's decision to proceed with planning the art museum. It was assumed that construction would immediately follow.

A Project Charge Committee was named,[7] with Shari Edelson joining me in representing the arboretum. The Project Charge Committee, a standard feature for very large capital projects, functioned similarly to the faculty and staff committees we assembled for arboretum projects. An executive committee was also named, but that group (which included me) never met. The president later informed me that Bill Sitzabee kept him apprised of progress. The two of them and Erin Coe probably guided major decisions.

The art museum plan involved two sequential projects. The first focused on verifying the program outlined in the ASG plan, validating space needs, and aligning the project scope with a budget that was becoming clearer as fundraising progressed. We hired the firm of DLR / Westlake Reed Leskosky for these tasks, which were completed around the end of February 2019.

As noted above, the arrangement of buildings now needed to be redesigned. DLR proposed four options for an art museum as the first building in a re-phased cultural district. One option was to locate the art museum at the top of the rise from Park Avenue, replacing the arboretum parking lot and placing the entire building north[8] of the eastern entrance to the H. O. Smith Botanic Gardens. However, space here was limited by underground utilities, and the building would need to extend over Bigler Road, severing its connection with Services Road. This was deemed unacceptable. All three other options placed the art museum entirely south of the arboretum entrance, maintaining direct access into the gardens from the existing arboretum parking lot and from the lot behind the Katz Building. This also positioned the art museum as close to campus as possible on the arboretum site, which seemed desirable.

The second project was an art museum building design based upon the program identified by DLR. For this project, Allied Works Architecture (AWA) was chosen following interviews with three firms in April 2019. The deciding argument for AWA was Erin's belief that selecting them, apparently a better-known firm in art museum circles, would help convince donors to give money to the project. We began meeting with AWA staff in May.

Unfortunately, planning the art museum was not the happy and heady task it could have been. Out of the dozen or so arboretum-related projects in which I was involved, this was the only one where tension was a dominant mood in our planning meetings. It's not that we never disagreed in other planning projects, but those disagreements were always resolved through compromise or concession in an amicable give-and-take. The art museum project was different.

I have mentioned in passing a number of personnel changes that had occurred during the previous couple of years, and I believe these had an indirect effect on the trajectory of the art museum project with respect to the arboretum. With hardly an exception, the OPP associate vice president and its administrative and professional staff on the Project Charge Committee had no history with the arboretum project and were mostly new to me. The art museum director was new, and Bobbi Korner had retired as dean of Arts and Architecture. In her participation in the Arboretum Steering Committee and the ASG cultural district project, Bobbi was always generous and open, a collegial and considerate partner, and typically an ally on matters regarding the arboretum. Then—just as we began determining the art museum's location and footprint in the summer of 2019—our highly capable project manager, Rachel Prinkey, was replaced when she left to take another job at the University.

The new colleagues were competent professionals, but there was a general lack of assertiveness within the group toward holding AWA responsible for meeting the project's goals. Yet this was precisely the Project Charge Committee's principal responsibility: "maintaining project scope in alignment with the University goals." Rachel reminded us of those goals in a July 19 email shortly before she left, and they included:

- "a new art museum that is integrated with the arboretum and does not detract from the arboretum. The new museum building should enhance the overall sense of place of the arboretum, while still retaining its own unique identity as an art museum,"
- "[enhancement of] the visibility, visitorship, and guest experience for both the art museum and the arboretum at Penn State," and
- "a planning and programming concept for the new art museum that delivers a fully functional, stand-alone museum, but also considers a future art museum expansion and connectivity to the broader cultural district plan."

After planning got underway, it soon seemed that AWA and its lead architect, Brad Cloepfil, had little interest in "integrating with" and "enhancing" the arboretum, a perception reinforced by statements he made that came back to me indirectly through other members of the Project Charge Committee. His interest was the art museum and its prominence and accessibility. Our directions to him and his team to modify their evolving plan for better alignment with the project goals were repeatedly omitted from AWA's meeting records and either ignored in plan revisions or acknowledged with the least possible meaningful change. This became a source of tension and distrust in our meetings.

The first four concept plans proposed by AWA, in July 2019, were sprawling, meandering complexes that would have taken up nearly all available space on the east side of the H. O. Smith Botanic Gardens. Two-story museum galleries were sited toward the north (over the existing parking lot and at the high point of the site), and one- and two-story educational/administrative wings were extended toward Park Avenue, past the "southern" edge of the walled gardens. Non-campus visitors would park in the Katz lot and access the botanic gardens *through* the museum. All of the options that we were presented at various times showed a possible future conservatory in the space intended for it since 2006, but none showed a planetarium, and only three provided for a STEM museum.

I was on vacation at the time of the July meeting, and AWA did not issue meeting minutes or notes. Shari sent me an email describing AWA's plans, though, and I quickly replied with alarm over their impact on the arboretum. I copied Rachel, and this was the occasion of her email reminding everyone of the project goals. She followed up with specific instructions for AWA to prove that space remained available for complete development of the cultural district, and she included, as a reminder, square footage targets for a STEM museum, planetarium, and conservatory. Together, these amounted to clear instructions to AWA, and indirectly the Project Charge Committee, to pursue *all* the goals in its charge.

AWA presented three new concept plans in mid-September. Only one indicated a location for the STEM museum, and none showed where a planetarium could be built. This was no mere oversight. The available land was limited, and other elements of the cultural district were simply ignored if their art museum scheme left no room. In contrast to AWA's concept plans, every location/footprint option offered by DLR in February provided space for all of the buildings intended for the cultural district, and they integrated well with the gardens by allowing a direct, on-axis entrance from the Katz parking lot. They also would have avoided thorny and contentious constraints caused by trying to find room for too much building in too small a space, which was limited to the south by access to the gardens and to the north by underground utilities.

As with the July plans, each of the new schemes crowded the museum galleries onto the high end of the site, away from campus. Cloepfil explained that his attraction to this location was that it afforded "picture window" views toward Hartley Wood and Bald Eagle Ridge. Indeed, we placed the arboretum's Overlook Pavilion at the top of the rise partly for the same reason. It complemented our focus on landscapes and nature. For an art museum, though, this seemed an odd priority given that the aesthetic attention of art museum visitors is indoors rather than out. Furthermore, the ridgeline location placed the galleries at their farthest possible distance from campus. However, a ridgeline location would give the museum a flattering prominence and enable its entrance to align directly with pedestrian access from the Katz parking lot.

Over the course of five or six work sessions, from July to September, sporadic challenges from Steve Watson, me, and others to the wisdom of putting the museum at or near the ridgeline, as AWA did in all seven of the options it proffered during that time, were evaded or dismissed. It became almost impossible to raise the question again, as if one were returning to a topic that had been thoroughly discussed, evaluated, and decided (but hadn't). Rachel Prinkey's summer departure as project manager had left a noticeable leadership vacuum.

Discussions in the September meeting led eventually to the choice of a "connector" option, which provided an east-west passage to the gardens between two building units, one for galleries and one for administration, joined overhead by a second-floor hallway. However, in a November meeting, AWA was still proposing to squeeze 70 percent of the building into 30 percent of the available space. This pushed the garden entrance awkwardly toward Park Avenue, even to the extent of intercepting the Strolling Garden walkway almost tangentially at about the "five o'clock" position (if the Overlook Pavilion is at twelve o'clock, or "Penn State north"). (Our existing, on-axis entrance was at three o'clock.) Steve Watson asked for justifications, and the Project Charge Committee demanded a design that preserved the three o'clock garden entrance by shifting some of the museum program to the south.

Once again, AWA provided no minutes for the November meeting, and their December meeting sketches failed to acknowledge our directives. Instead, they shifted space for events to the north unit, which placed the garden entrance even more off-axis in at least two concept sketches. We pushed back, and in January 2020, we homed in on a configuration that seemed the best possible, even though it did not provide a highly desirable direct entrance into the gardens from the Katz lot. The architect rarely attended our meetings, and some of us developed the impression that those acting under his direction knew the plans would be objectionable and were uncomfortable bringing them to us. I was disappointed that so little was done to control, or perhaps to honor, the University's formal planning process. This was a significant departure from all of my previous experiences.

Construction began in the summer of 2021 and was completed in the summer of 2023 at a cost of $85 million. The arboretum staff occupied their offices in July 2023, and the museum opened to the public in June 2024. The building is handsome inside and out, and the museum staff and patrons should be pleased.

I had hoped for a building with more grace, but that wistful vision, prompted by the Gund plan, was not essential to our goals for this project. However, it was intended—directed, even—that the building integrate with and enhance the arboretum, and this building does not. Rather than being designed for the arboretum, the building seems merely to fit within the available space. With only minor changes of a practical sort, it could be equally suited to empty lots on dozens of university campuses. It was understood from the beginning, of course, that making room for a cultural district would to some extent diminish the H. O. Smith Botanic Gardens, if only by requiring the gardens to give up space. The bargain was to create something that would be greater than the sum of its parts. That was the vision that President Barron presented to me. Unfortunately, Brad Cloepfil's design falls short of that vision. If the rest of the cultural district can ever be built, it certainly cannot be done without a far greater impact on the arboretum than contemplated in the now-obsolete ASG plan or DLR's recommended siting options.

My only enduring disappointment is that, despite our insistence, the entrance to the gardens is not aligned with its dominant features, and the route from the Katz parking lot is circuitous. For visitors arriving by vehicle, the gardens are in the museum's backyard. Paved, pedestrian egress from the lot is directed toward the museum entrance, and there is no simple route to the gardens. Returning visitors tend to avoid the crosswalk and cut diagonally across the road, and new visitors are sometimes confused about where to go next after finding themselves at the museum entrance. The indirect route into the gardens must seem a nuisance to our visitors, who pay this toll of annoyance on each visit.[9] It could be partly rectified with a new walk on either the Katz side or the museum side of the road, and perhaps someday it will be.

The Childhood's Gate Children's Garden

Schematic Design

Our pursuit of a home for arboretum operations has carried this narrative beyond events yet to be described. To resume a chronicle of the growth of the H. O. Smith Botanic Gardens, we must return to a time when the initial construction was still underway.

Phase I was designed to occupy and landscape as much of the Mitchell Tract as possible with the funding available. However, expenditures for Phase I were only a small fraction of the total anticipated cost of the Mitchell Tract plan, and most of the landscaping involved only grading and seeding. Remaining parts of the plan would be built piecemeal as we raised more money, and to interest donors we needed preliminary designs and graphic renderings. The obvious path forward was to pursue the next feature with the highest donor appeal and hope that the more we completed, the easier it would become to find funding for what was left.

An early piece of advice from Missy Marshall was that children's gardens were popular, and she included one in every version of her master plan for the Mitchell Tract beginning in 2002, although it had a slightly different label until 2008. At that time, gardens designed intentionally for children were a novelty in the world of botanic gardens. The trend seems to have started in 1993 with a garden that opened on the Michigan State University campus, and I am aware of only three that were created before 2000.

I visited many children's gardens beginning in 2000. They differed quite a bit in character. Even discounting for a one-day visit that might not be representative, visitation in some gardens seemed surprisingly sparse. I wondered why this was so, which led me to consider what we would *not* want in a children's garden. Prominent among these (and present in many gardens) were playground equipment, sculptures of fanciful creatures, and signs and demonstrations imparting simple lessons, which were often present in gardens that were otherwise plain and sterile. I never saw a garden where playground equipment was the most popular attraction for children, although that appeared to be the intent. It's not that I believed a children's garden should not encourage play, excite fantasy, or offer educational opportunities. On the contrary, I believed that the richest forms of play do not require equipment, that fantasy requires only imagination and the opportunity to exercise it, and that didactic signage is soon ignored. It seemed better to aim for a garden replete with mystery, wonder, and beauty for children to enjoy on their own in imaginative play—and with latent educational content that talented teachers could tap. We wanted a garden that rewarded repeated visits, which some of those I visited probably did not. I wanted a garden full of opportunity for those rare childhood experiences, often buried in memory, that plant the seed for a lifelong interest, enthusiasm, or even vocation. However, I had no idea what such a special garden might look like.

In mid-2008, as construction began on Phase I, it appeared that we could pay for a children's garden schematic design by bundling our unrestricted gift receipts (which had taken a significant jump) with proceeds from the Marcia Day endowment mentioned earlier. We required a rough design, a few rendered vignettes of garden scenes, and an estimate of probable cost. The garden, as outlined in the master plan (appendix I), would be just under an acre in size. We had about $64,000 to spend on preliminary design work.

Kate Reeder and I met with Judy Larkin, Gordon Turow, and Dave Zehngut in August 2008 to plan the next steps. Judy would develop a list of firms with relevant experience, and we would collaborate on a request for proposals. We discussed the composition of a project oversight team, which was soon organized.[1] I particularly wanted to include Linda Duerr, director of the Child Development Laboratory in the College of Health and Human Development.[2] Linda had visited me in April of that year to discuss opportunities in the arboretum for preschool children at the soon-to-be-built Hort Woods facility. She wanted to incorporate an environmental education component into their curriculum for preschoolers. I was impressed by her personality, intelligence, and suggestions for attracting children to the gardens. We eventually hired Linda to develop and oversee educational programs in the Children's Garden.

Finding firms with a track record in the design of children's gardens proved difficult. We learned this was a niche market for design firms, which seldom featured children's gardens on their websites or in advertising. However, we eventually identified a respectable list of candidate firms and sent them requests for a proposal in February 2009. The letter described our vision of creating a "delightful, interactive, learning environment for small children, aged 11 and under, and their families. [A design that would] include spaces where children can be children . . . and safely discover the wonderful world of plants . . . not a playground, but a contemporary educational garden." We did not quite say it in the letter, but as Gordon Turow put it later, our goal was to "set the national standard for children's gardens."

We interviewed four firms and selected EDAW for the work. EDAW (which became part of the engineering and design firm AECOM during the contract) was a full-service design firm. For our project, they partnered with Emmanuel Didier, an independent designer in Fort Collins, Colorado, who was a protégé of one of the first designers of children's gardens, Herb Schaal. Didier and Schaal had worked together on the just-completed Downing Children's Garden in Wichita, and we were impressed with the garden director's enthusiasm for Emmanuel's talents. Our contract was with EDAW, but the design that emerged was Emmanuel's work. (He subsequently founded his own firm, Didier Design Studio.)

The outlines of the design were identified in three-day workshops and charettes on campus in October and November 2009. Earlier, I mentioned Emmanuel's skill at quickly developing compelling design concepts after probing our values and goals. Following two visits and iterative meetings with the oversight team and a stakeholder group,[3] we settled on a scheme that would evoke the regional ridge-valley-ridge landscape and its natural and cultural history. Elements of the design could be developed around land use patterns, history, geology, and native plants and animals in ways that would be both visually attractive and rich in educational content.

The schematic design (appendix M) with narrative text was finalized in March 2010 following a third on-campus meeting in January and written feedback from the oversight team and arboretum staff. Especially encouraging were Linda Duerr's thorough review of the plan and the comments she received from early childhood teachers with whom she had shared the plans:

> Wow! (the most common comment)
>
> I can see myself as an adult loving to visit this space again and again.
>
> It seems that there are . . . many little attractions that will create the need for many more visits.
>
> This concept transcends all ages. I can see coming to visit with my nieces and nephews and with my grandfather and we will all learn something.
>
> It is amazing and not just from a child's view but from an adult perspective. I want to go there by myself!

We had deliberately targeted children, of course, but reminded Emmanuel that there were over forty thousand post-adolescents just across the street. The unprompted responses from Linda's colleagues made me confident we were on a path toward a garden with universal appeal. Incidentally, one of the unexpected attractions of the Children's Garden is the children themselves, whose pleasure in the garden seems to provide vicarious enjoyment to many adults and perhaps especially to elderly visitors.

Fundraising

The EDAW team estimated that design and construction of the Children's Garden would cost about $2.5 million, and we calculated an endowment need of $1.7 million to pay for maintenance and a part-time educator. We would need at least one seven-figure gift. In January 2010, Rod Kirsch met with an alumna who expressed interest in the Children's Garden and could make a gift of that size. If a donor were to give us $2.5 million, the Marcia Day endowment would cover part of the annual operating costs, leaving about $50,000 still to be found. Additionally, we had about $2.5 million in funding opportunities for named features such as pools at the entrance, vegetable plots, giant toadstools (for seating), a cave, and a prominent glass house. We were optimistic.

Nothing had changed by June 2010, when Rod Kirsch and I met with Rod Erickson to see how much latitude we had for proceeding without all of the funding in hand. After giving Provost Erickson an outline of the numbers, he offered to fund a shortfall in operating costs of $50,000 annually if we could get a $2.5 million gift. Rod revealed to Patrick Williams and me in July that he had been talking with Helen and Ed Hintz about funding the Children's Garden. Helen was especially keen on the idea of an educational and fun garden that could attract joint visits from children and their parents or grandparents, and she eventually visited children's gardens in several cities to learn more about them. Rod proposed that she and Ed make a gift of $2.6 million to get the project underway. He asked if I could meet with the Hintzes when they were next in town to discuss our vision and the details of the project.

Helen, Ed, and I met for about an hour at the Overlook Pavilion on September 17, 2010. Afterward, I wrote an email to them that contained the following paragraph, which I include here because it summarizes our goal for the Children's Garden:

> Our greatest desire is to create a Children's Garden that will influence lives in a positive way. Education is the vehicle for that, but it's been my experience that transformative moments are tied to subtle, fragmentary, and even fleeting qualities of what is said, who is saying it and how, and the environment in which it is said. No one can set that up. But I think we can make those moments possible by building a garden with a richness of appealing features and staffing it with good people, and that is our objective.

I thought our meeting went well, and Ed made a similar comment to Rod. Rod was optimistic that they would make the gift, and he received verbal confirmation a few months later that they would donate the amount we requested. The Hintzes are a wonderful couple. They have been very generous to the University, and they always seemed more interested in the good of the University than in personal recognition. As they later expressed in an interview, their motivation for the gift was to inspire other alumni to join them in "shouting for Penn State and for this arboretum from the rooftops . . . to build the University into the best that it can possibly become."

In June 2011, Rod Kirsch, Dave Lieb, and I met with Skip Smith to explain where we were with planning and funding the Children's Garden, and we asked him for a gift of $1.5 million to endow maintenance and operations for the garden. He showed interest but did not commit. We met again in August, and he asked that we make some changes to the proposed gift agreement. In particular, he wanted assurances that the University would never charge an admission fee to enter the arboretum. We changed the language to make this policy commitment unless a fee became necessary for the financial health and physical well-being of the H. O. Smith Botanic Gardens. Skip signed the gift pledge on August 31. A December press release announced the Hintz and Smith gifts,[4] using the words "childhood's gate" from the University's alma mater to refer to the garden. These words would eventually be chosen for the garden's name.[5]

Final Design and Construction

The Hintz and Smith gifts, the Marcia Day endowment, and the prospect of more donations gave us what we needed to proceed. We went to work immediately after Skip signed the pledge, and within three days the wheels were in motion to start the project with AECOM (formerly EDAW).

The plan was to complete the design over the winter and begin construction in spring 2012. To this end, AECOM conducted two intensive, two-day design meetings in October and December. In addition to the oversight team, these meetings involved nearly seventy-five faculty, staff, and community representatives[6] participating in nine workshops (sustainability, catering, art opportunities, geology, native plants, and so on). Some discussions were lively, and the designers were delighted with their input. Garden details quickly fell into place as a product of Emmanuel's design genius and his skill in extracting helpful guidance from our large "stakeholder" group.

The bid package was issued to contractors in mid-April with a deadline of May 10 for submitting bids. The deadline was extended twice—ultimately to June 30—after mistakes and inconsistencies were discovered in the bid documents. These issues were causing frustration among contractors as they prepared their bids. The fault lay with the old EDAW team that had developed the construction drawings and engineering specifications. We secured a commitment from AECOM to bring in new staff, but we remained concerned that the confusion was making contractors overly cautious as they estimated their costs. Ultimately, we put the bidding process on indefinite hold after discovering additional errors, including a simple spreadsheet error that led the planners to understate the project's expected cost by $320,000. In July, an independent constructability review commissioned by OPP projected a construction cost that was $660,000 higher than our available funding.

Working through this with the new AECOM team was a chore, and we were forced to make money-saving adjustments to the project scope. Corrected bid documents were finally issued with a March 2013 bid deadline. We awarded the contract to Leonard S. Fiore, Inc., and construction commenced in mid-April. The project's projected and final cost was $3.4 million, and we managed to cover all expenses with the original and additional gifts.

We had a delightful groundbreaking event on May 30, 2013, with over 220 guests. Near the end of the speeches,[7] about twenty children from the Child Care Center at Hort Woods and the Bennett Family Center, ranging from three to five years old, danced to the front of the audience, led by a flutist, and sang the "rainbow song." Linda's young students had painted designs on our shovels and the "sandboxes" that held the ceremonial soil. Penn State Hospitality Services served up an astonishing array of special snacks: peanut butter and other sandwiches cut into the shapes of flowers and butterflies (no crusts!), "garden" pasta salad in miniature buckets, mousse flowerpots with sugar flowers, "dirt" cupcakes with gummy worms, and much more, all beautifully executed and accompanied by stunning decorations. Rod Kirsch called it one of the best "feel-good" events he had ever attended, and I had no reason to doubt him.

One of the underappreciated features of the Children's Garden is its extensive use of stone and the beauty of several pieces. The large "sliced-bread" stones at the entrance—intended to evoke the profile of our surrounding mountains—were cut from

Curwensville, Pennsylvania, sandstone by the Russell Stone Products quarry, as were the stones laid in arched courses near the cave to represent the folded bedrock in our region. This is the same stone used in the construction of Phase I and, later, the art museum. Emmanuel's design included specifications for eighty-three naturally weathered limestone and sandstone boulders, ranging from 2.5 to 10 feet long, to be found and positioned according to plan. Derek Kalp found most of the limestone boulders by searching University lands, and some came from excavations for a HUB expansion project. Derek and Judy Larkin selected sandstone boulders at the quarry based on size, shape, texture, and occasionally their colorful, mineralized surfaces.

Stonemason Ron Bair and his sons, Jeremy and Ronnie, did the "folded bedrock" stonework and delicate placement of the "sliced bread" sandstone slabs. We learned from Fiore crew members that the process of cutting and fitting the stones was taking far longer than the Bairs had expected. They were losing money, but they neither complained nor compromised on their careful craftsmanship. (When the work was done, after making inquiries through Judy and the Fiore supervisor, I arranged through OPP to give an appropriate bonus to the Bairs.) I do not know whether everyone on the job was as conscientious as the Bairs, but the entire crew seemed to appreciate the project's special character and take pride in their involvement. To show our appreciation for their work, we invited the tradesmen and supervisors for a picnic in the Children's Garden on a Sunday in August, after the garden was completed. Most came, accompanied by their families, and we enjoyed sharing our gratitude and their pride in a job well done.

Construction was essentially completed on June 30, 2014. We conducted media interviews in the garden on July 3 and opened the garden to the public on July 7. I was in Utah that day, but Shari Edelson reported a steady stream of visitors totaling many hundreds, and the parking lot remained full throughout the day. Here is Linda Duerr's report to me, slightly edited for brevity:

> What a wondrous day. Busy, fun, challenging, amazing, fulfilling . . . to finally be in that garden WITH children . . . a gift. There were so many wonderful moments, funny things that kids said and did . . . I heard only positive comments, adults were thrilled, the faces of the children were really the best story. Lots of photos taken and hope they will make their way around. I would say the cave and the creek were the standout favorites. There were several families who came in the morning and in the afternoon. There were elderly couples walking around and smiling, PSU students towards the end of the day but mostly families and an age range from a 2-week-old to our 98-year-old volunteer, Grandma Norene!

The dedication ceremony took place on a Sunday morning, September 7, 2014.[8] Hospitality Services once again provided us with whimsical, children- and garden-themed food items, different from those prepared for the groundbreaking but equally clever and delightful. There were approximately four hundred attendees, and they seemed to take a great deal of pleasure in what had been created from Emmanuel's design.

In 2016, we were contacted by Lolly Tai, a professor of landscape architecture at Temple University, who was gathering material for a book on children's gardens. We provided what she needed, and the Childhood's Gate Children's Garden became one of nineteen children's gardens featured in her book *The Magic of Children's Gardens*, published in 2017. Although the Children's Garden is only a small portion of the H. O. Smith Botanic Gardens, it has become one of the most-visited features in the arboretum.

The Bison Sculpture

As we were finishing the Children's Garden, Penn State's "For the Future" capital campaign was also coming to an end. Peter Tombros, the campaign chair, approached us about the possibility of a suitable gift to the Children's Garden in honor of Rod Kirsch's leadership of the campaign. The Tombroses and others of Rod's friends and admirers would sponsor the gift.

Because Rod grew up on a ranch in western North Dakota, the donors agreed to fund the creation of a full-sized bronze sculpture of an American bison. The Children's Garden features many bronze sculptures of animals that are native to Pennsylvania, and bison once roamed our area before their extirpation (probably in the late eighteenth century). I suggested J. Clayton Bright as a sculptor, having seen his wonderful sculpture of a reclining cow on the grounds of the Brandywine Museum of Art.

Clayton presented us with a fee proposal in February 2015 that was within our budget, and we got permission to purchase his services without going through a bidding process. With our commitment to move forward, Clayton produced some working sketches, which were successfully submitted for approval by the University's Public Art Committee.

As it happened, Clayton had a neighbor who raised bison for the market. This gave him the freedom and convenience to observe and sketch living animals as needed, and he obtained the head of a harvested animal to use as a model during the remaining weeks of winter in his (unheated) studio. To fit within the available space, Clayton's composite "subject" was a two- to three-year-old animal and not quite as large as bison can become. However, to most eyes, the sculpture appears full-sized, and it reportedly caused quite a few stares when Clayton brought it to us from Chester County, Pennsylvania, in the bed of his pickup truck. It was installed in the Children's Garden on September 30, 2015. On October 11, a dedication ceremony was held in Rod's honor before about sixty people who had figured importantly in the campaign's success in raising $2.2 billion. The bison occupies a prominent location near the entrance to the garden.

An Education Coordinator Is Hired

As we worked with AECOM in 2012 to resolve issues in the construction document package, Linda Duerr and I began discussing how to implement educational programs in the Children's Garden after construction. Helen Hintz was interested in this, too, and the three of us met in February 2013.

During our conversation, Linda hesitantly suggested splitting her position in the College of Health and Human Development to serve half-time with the arboretum as our coordinator of children's educational programs. Helen and I thought highly of Linda and were thrilled with this unexpected offer to work for the arboretum. I promised to investigate the possibility while noting that I would not have enough funding to hire her until the 2015 fiscal year. Helen said that she and Ed could help financially to manage a quicker hire if one could be arranged, and I relayed this information to Rod Kirsch as Helen requested.

Linda and I met with her department head, Douglas Coatsworth, in April 2013 to discuss our desire to split her position. Rod and I followed this by meeting with their dean, Nan Crouter. An agreement was reached, and Linda began her part-time position with the arboretum in August 2013. She and Helen developed a friendship and often met when the Hintzes were in town. Helen continued to support Linda's work by providing funds to hire student interns. Although the other half of Linda's position was eventually moved to the College of Education, she continued to serve as the arboretum's half-time coordinator of children's educational programs until, for medical reasons, she took a year's leave of absence beginning in June 2019 and then retired. Under her leadership, the Children's Garden blossomed as a place of enjoyment and discovery for children and their families. Sadly, Linda passed away in August 2024. Rebecca Horwitt helped fill in during Linda's absence, and we hired Jennifer Hooven to become Linda's full-time replacement in July 2021.

The Pollinator and Bird Garden

As proud as we are of the Children's Garden, our finest garden *might* be the Pollinator and Bird Garden, although the journey there took thirteen years over a path strewn with false starts, a global pandemic, and a protracted quest for funding.

A "butterfly" garden in the arboretum was urged as early as the stakeholder meetings for the Sasaki plan in the late 1990s. Robert Snetsinger, professor emeritus in the Department of Entomology and a member of the planning oversight team, was a particularly strong advocate. However, a pollinator garden did not appear by name in our plans until the 2008 revision of the Mitchell Tract master plan (appendix I). Phase I construction included about half of this planned garden, which was located between the Rose and Fragrance Garden and the Esplanade. Some will remember its dominant features as two circular beds enclosed by walkways.

The Phase I pollinator garden had some charm, particularly in late summer, but it was too small and not well executed. The plants grew too tall for the beds to be viewed across, as Missy Marshall had intended (she did not select the plants). Similar plants were alongside the walkways, so visitors were literally "in the weeds" and could have little sense of the entire garden. We put a couple of Adirondack chairs in a back corner, and the garden became somewhat popular as a secluded space to rest or study. But it was less often visited than traversed on the way to somewhere else.

The Oehme, van Sweden Plan

Just as Phase I was completed, in early fall 2009, two new hires in the Department of Entomology—Christina Grozinger and Harland Patch—were forming a faculty interest group on the topic of pollinator biology. I was invited in my capacity as a faculty member. While declining to join, I suggested to Christina that their group adopt our still-unfinished pollinator garden and help steer its future development. As it happened, she and Harland had already toured the H. O. Smith Botanic Gardens with Rick Johnson and were eager to collaborate. I met with their faculty group to brief them about the arboretum and exchange ideas about using and improving the pollinator garden. Harland offered to write a new planting plan, and the group agreed to work together to design informational signage. Both tasks were completed by early 2011.

Our conversations eventually converged on the idea of an entirely new and ambitious pollinator garden that would be distinctive in the world of botanic gardens. In July 2010, Penn State had hosted the inaugural International Conference on Pollinator Biology, Health, and Policy. Members of our faculty were prominent within this community of scientists, and Harland was knowledgeable about what was being done internationally in pollinator garden design. We were excited by the prospect of creating an exemplary garden—perhaps the "best," if such a thing could be measured. As always, to raise money for such a project, we would need a schematic design and an estimate of the probable construction cost.

Our initial idea was that this new garden would occupy the Pollinators' Garden space in the 2008 master plan, about two-thirds of an acre. At some point, however, we expanded the footprint to almost two acres by replacing most of the still-unbuilt Demonstration Gardens arrayed around the Esplanade in the master plan. Those planned gardens had never attracted much interest or comment, and I had little hope of ever raising the money to build them. Raising money for the pollinator garden, however, seemed quite possible.

By April 2012, with pledges of support from Rod Kirsch, Bruce McPheron (dean, College of Agricultural Sciences), and the Department of Entomology, I was able to gather enough funds to pay for a schematic design project. With Judy Larkin and others in OPP, we prepared and issued a request for proposals and assembled a project oversight team.[1] The contract was awarded in June to the firm of Oehme, van Sweden & Associates (OvS).

The kickoff meeting for the project took place in September 2012. Harland gave a presentation on the biology of pollinators, their plant and habitat requirements, and possible garden themes, and we engaged in a lively discussion. In a subsequent email exchange, I asked whether it would be feasible to design a garden that would attract *all* of the hundreds of pollinating insect species native to central Pennsylvania. We discussed this goal in the next workshop, held in December, and it became the central idea of all subsequent planning leading up to the final garden installation in 2021. It was a daunting idea, especially since no such comprehensive list of insects exists, and it appears to have been completely novel for any region of the world.

OvS presented their final plan (appendix N) in a June 2013 conference-call meeting. They estimated a cost of $3.9 to $4.9 million for the garden they proposed. Unfortunately, this last meeting only confirmed what we, especially Gordon Turow and I, had already privately concluded—that the OvS design was not what we wanted. Their site plan imposed an unnatural circulation pattern that integrated poorly with the existing gardens. This objection was not new, and the design could have been reshaped to fit our needs over the months since our first meeting (at least we believed so). However, the

OvS team resisted any reconsideration of their initial scheme. In their defense, OvS had trimmed their fees substantially to meet our budget, and they may have wished to avoid losing money. We were looking for someone else to take on the project even as we concluded this last meeting.

The M·T·R Plan

The OvS project was not entirely a loss. We had a footprint and a construction cost estimate that would serve for a garden of comparable size and complexity. Through Judy, I began discussing a redesign with a couple of firms that were familiar to us. The idea was to start with the OvS report but to sketch an entirely new site plan and provide a few watercolor-like renderings for fundraising purposes. We settled once again on Missy Marshall (M·T·R Landscape Architects), confident that her critical eye and creative mind would lead to a site plan that appealed to us. Her price was reasonable, and the Office of University Development provided the funding.

After working through preliminary design ideas, Missy suggested adding an adjacent bird garden. As well as providing the usual nesting and foraging habitat, the bird garden she envisioned would be designed to support the insect populations that birds (even frugivores) need to raise their young.[2] This dovetailed thematically with the pollinator garden, and we added a bird garden to her contract at an additional cost.[3] To assist Missy with the pollinator garden, Harland and I wrote a three-page narrative in January 2014 discussing the weaknesses of existing gardens, restating our lofty goal of attracting all pollinators native to the region, and describing the characteristics and components of a garden that could achieve our goal. She presented a conceptual design with renderings for a pollinator garden in March 2014 and a bird garden in December 2015. Both products informed the garden that was finally built, but the pollinator garden design (appendix O) was especially influential. Missy also provided site narratives and revised cost estimates.

Throughout this time, Patrick Williams and I worked with Rod Kirsch, Michael Degenhart (assistant vice president of gift planning), and others in Old Main to get the donations needed to build two gardens, one for insect pollinators and one for birds. Some of this involved complicated real estate gifts from Skip Smith that took a long time to finalize. The two gardens were the subjects of our annual fundraising appeals between 2014 and 2017. Rod and I met with Provost Nick Jones in April 2015 to explain where we stood and to ask whether he would temporarily backstop the arboretum on maintenance expenses (about $116,000 annually) if we could raise enough money to begin construction. The provost agreed. By early January 2018, we had accumulated nearly $6 million in gifts from Skip Smith, his sons, Galen Dreibelbis, and numerous smaller donors. I met again with Nick Jones and Kathy Bieschke (vice provost for faculty affairs and now chair of the Arboretum Steering Committee) to reaffirm that Nick would temporarily cover maintenance costs, if necessary. He again agreed, and this was enough to begin final construction plans.

The Didier Design Studio Plan

The M·T·R designs placed the pollinator and bird gardens next to each other, and we decided to hire one firm to develop the final designs and construction documents for a single, integrated Pollinator and Bird Garden. After interviewing three shortlist firms for the final design project, we selected Didier Design Studio in early May 2018. This was Emmanuel Didier's new firm, now separate from AECOM. The selection was somewhat unusual because Emmanuel had never designed a garden like this, and he had little biological expertise. However, we had a high opinion of his design skills from the Children's Garden project, and he teamed up with Phyto Studio (represented by Claudia West and Thomas Rainer) for biological guidance. Theirs was a well-known firm specializing in "ecological horticulture." Additionally, we assembled a project oversight team[4] populated with entomologists, ornithologists, plant scientists, and ecologists from the University, and I was confident that most of them would bring enthusiasm to the project. If Emmanuel could deliver a good design, our collective efforts could guide the composition and arrangement of plant collections.

We held two-day planning meetings and site visits on four occasions between August and December 2018. The site plan was essentially completed after November (appendix P), and the intricate details—soil selection, grading, plant selection, and physical structures—were worked out over the next seven months. It challenged our collective knowledge to identify the plants needed to attract hundreds of species of insect pollinators. Some pollinate many species while others pollinate very few; the latter, of course, are more challenging to accommodate. Because the host preferences for some insects are largely unknown, we did not expect to immediately reach our goal of attracting all of the pollinators in the region. Accordingly, we included space for research trials to guide the future selection of additional plants after the garden was built.[5] The final plan was to install 143,000 plants belonging to nearly three hundred species. To support this diversity of vegetation, six soil types were prepared or imported, one of which was partially saturated with water. The plan thus featured seven edaphic zones (see appendix P).

We were attempting to replicate a complex system of overlapping mutualistic relationships between two arrays of species: one of plants and one of insects. This was not easy, and we faced an additional challenge in accommodating the special habitat needs of some insects and birds for nesting purposes. The key collaborators in this effort were Harland Patch, Shari Edelson, Margaret Brittingham (professor of wildlife resources), and Claudia West. Using information provided by the others, Claudia identified sixty-five thematic "stories" about plant/bird/pollinator relationships that could be illustrated in the garden through the design and composition of vegetation (a detail is shown in appendix P).

In July 2019, we contracted with Leonard S. Fiore, Inc., to build the Pollinator and Bird Garden through OPP's Job Order Contract (JOC) process. This was different from the usual bidding process, and it took several months to finalize the cost details. However, we chose to use the JOC process because it gave us joint control with Fiore over the selection of a subcontractor for the all-important landscaping work. We had been unhappy with the landscaping firm hired for the Children's Garden, and we wanted to avoid a similar experience in this new project. Scott's Landscaping became the "best value" bidder for the landscaping subcontract.

Construction began in October under an "early start" contract while we continued to work out costs with the Fiore team. Most of this burdensome and meticulous task fell to Jesse Wells, facilities contract specialist in OPP, and I am grateful for his good work in the service of our desire to control the landscaping subcontract. It was a hectic time for the rest of us, too, especially Judy Larkin (as project manager) and Shari Edelson. Shari was relieving me of much of the usual burden of monitoring and supervising details. In addition, to save money on the contract, she volunteered to source and order over $400,000 worth of herbaceous perennial plants to be delivered before the fall 2020 and spring 2021 planting seasons. Derek Kalp took on the job of assembling the more than three hundred tons of limestone boulders specified by Emmanuel as landscape accents. Some were purchased, but Derek and Lake Graboski, a graduate student, spent many days on a "boulder hunt," scouring remote Penn State properties for the most interesting ones. Derek's term for these uniquely weathered relics was "goonies."

Two cases of COVID-19 in Pennsylvania were announced on March 6, 2020, setting off a year of fear and disruption. On March 19, Governor Tom Wolf ordered all "non-essential" businesses to close, and construction on the Pollinator and Bird Garden was necessarily suspended. Technical service personnel were furloughed, and the University directed the rest of us to work from home unless otherwise necessary. It was not strictly necessary for me to be at the office, but it was easier to continue working there as usual. Even if COVID-19 were as contagious and deadly as initially feared, my office was safe, because only one other person ever seemed to be present in the entire building. Ironically, we were the two oldest faculty members in the department and (in our seventies) members of the most vulnerable cohort. Some might have figured we had no better place to be, but I've known my colleague for decades, and I'm sure we both would have done the same in our thirties or forties.

The arboretum could not be physically closed, but we locked the restrooms and posted on our web page that parking regulations would not be enforced. Some people came, of course, despite a stay-at-home order (later ruled unconstitutional) that had been issued by the governor on April 1. Even though construction was halted, daily exchanges of emails about matters to be resolved between contractors and us continued throughout the shutdown, so the project was not entirely at a standstill.

Incidentally, the shutdown came just as two new staff members joined the arboretum team. Tim Gould began working in February as our first dedicated manager of public programs, and Mandy DiBella started in March in a position that had been vacant since 2019, when Alisha Barton left to start a family. The shutdown made it unusually challenging for them to begin their new jobs, especially given the arboretum's public-facing mission and the nature of what they were being paid to do.

The governor lifted his suspension of construction activities on May 1, 2020, and the crews returned to the job site. However, most of the University staff remained on furlough or continued working from home, and many restrictions were still in effect, such as limitations on the size of gatherings and required masking. Through special pleading, we managed to have maintenance services for the existing gardens resumed in June. We needed this badly and were pleased to see our OPP maintenance staff relieved of their unpaid furloughs. (Shari had inadvertently forced the issue when she caused a union grievance by beginning to pull the weeds that were overtaking the gardens.)

With COVID-19 restrictions still in place, we were required to get special permission from Old Main to use volunteers, as had been planned, for the fall 2020 and spring 2021 installations of Shari's plant orders. Over the two seasons, 621 volunteers logged 1,863 service hours planting herbaceous perennials in the new garden, saving us a great deal of money. The final construction details were completed in June 2021. The total cost of the garden was nearly $9 million, including $4 million for the operating endowment. All of this money was raised through philanthropy, and we did not need to draw upon the provost's commitment to temporarily backstop maintenance costs.

We dedicated the garden late in the afternoon on Friday, October 1, 2021, before an audience of over three hundred. I had retired at the end of June but was graciously asked to speak at the dedication, accompanied by Rich Bundy (the new vice president for development and alumni relations), President Barron, Shari Edelson, Christina Grozinger, Margaret Brittingham, and Sanford Smith (interim director of the arboretum). I have already boasted about this 3.5-acre garden, so I will not elaborate further except to call it out as an excellent example of what can be achieved when designers and scientists collaborate in a unique design project. In 2024, the Pollinator and Bird Garden was the first prize Jury Winner in the Sustainable Landscape Planning category of entries in the Twelfth Annual Architizer A+Awards. Architizer is a large international awards program in architecture and landscape architecture, so this was a prize of some distinction. A map of this garden and the Palmer Museum of Art, in context with the rest of the H. O. Smith Botanic Gardens, is provided in appendix Q.

Engagement with the Center for Pollinator Research continued after the completion of the Pollinator and Bird Garden, and research and outreach activities in the garden began immediately. Christina Grozinger has maintained her involvement, and Harland Patch accepted an affiliate position with the arboretum as director of pollinator programming. Harland and Christina's department, Entomology, hired Heather Frantz in the summer of 2021 as their "pollinator educator," responsible for connecting entomology research to the arboretum and developing educational and interpretive resources for garden visitors. Coincidentally, Heather is the granddaughter of my earliest faculty colleague and mentor, Professor Henry Gerhold. I had no role in hiring her, but this seemed to close a personal circle that began in 1974 when Henry chaired the search committee that recommended my hire.

Transitions

As 2019 drew to a close, with construction on the Pollinator and Bird Garden underway and planning for the art museum nearly finished, I began to think seriously about retirement. Earlier in the year, in conversations with my dean and department head, we resolved a concern I had about the future of an academic employee I supervised. All that remained was to address the arboretum, which I wanted to leave in sound condition and on a promising trajectory. That meant preparing the way for a successor with the capability and freedom to elevate the arboretum to the next level of excellence.

The most vexing problem with transitioning to new leadership was that there was no full-time University position for an arboretum director, and the arboretum budget could not support one. My actual job at the University was a standard thirty-six-week faculty appointment with pressing responsibilities to teach, research, and publish. My role as arboretum director merely acknowledged extra responsibilities for which I was paid the equivalent of twelve weeks of salary, or ten hours per week in year-round employment. In fact, the arboretum work demanded much more time than that, which I found by working longer hours. Perhaps other faculty members would have taken the position on terms like mine, but the pool of candidates who would be both interested and qualified was surely small. A nationwide search offered better assurance of making the best possible hire, but what academic department would agree to fill a faculty vacancy with someone chosen partly for their qualifications as an arboretum director?

One solution was to find a donor to endow all or most of the director's salary, making it easier to secure an academic home for him or her at the University. I asked Patrick Williams to mention this idea informally to potential donors, and I discussed it with Kathy Bieschke during a phone call in July 2018. She called back the next day to say that the provost agreed that an endowed directorship with a fixed-term faculty appointment would be ideal, especially if the academic home would contribute part of the salary. He also agreed with my desire to fill the position through a national search.

This seemed to settle the question of how to fund and fill the position if we could find a donor, and we eventually did.[1] However, the administrative home of the arboretum remained an open question, and I wanted to do what I could to settle this to the best advantage for the arboretum. Was the existing Arboretum Steering Committee structure still the most desirable? It had worked well, but would it endure through the transition? It seemed unlikely that the next director would be granted the same level of independence that I enjoyed, which made the steering committee model workable. What were the alternatives? At least two deans would have happily acquired the arboretum as part of their colleges, but this would have violated our long-standing principle that the arboretum should be a University-level facility. Matters were in flux, especially with the impending relocation of the art museum to the arboretum, so there was good reason to worry a bit.

In January 2020, I surveyed the directors at twenty-eight university gardens and arboreta to learn more about their administrative arrangements. Most reported to someone within central administration (the arrangement recommended by our Sasaki consultants back in 1999), and this was especially true at gardens and arboreta that I regarded as our peer group. Exactly half of the directors had faculty appointments, and faculty appointments predominated at peer institutions. In April, I queried each member of the Arboretum Steering Committee on this topic through telephone calls. They expressed a variety of opinions. Most seemed to prefer the steering committee model while acknowledging the advantages of reporting to someone within central administration. There was strong support for a national search and hiring someone qualified for a faculty appointment in a relevant academic department.

I brought this information to a telephone meeting with Nick Jones and Kathy Bieschke on April 30, 2020. I suggested that the directors of the arboretum and art museum should share a reporting line to someone within the central administration. From our conversations, I knew Erin Coe favored this idea and was disappointed that our art museum was the only one in the Big Ten administered by a college. Bobbi Korner had told me before she retired that the art museum should be moved to central administration. Steve Carpenter (Bobbi's replacement as dean and Erin's new boss) appeared to share this opinion when we spoke earlier in April. Whatever their reasons for shifting the art museum to central administration, my preference for a common reporting line to central administration arose from the practicalities of occupying the same building, using some of the same rooms, perhaps cooperating on event rentals, and sharing a landscape in which both the arboretum and the museum had very high interest. (The museum maintains an outdoor art exhibit.) The arboretum and museum were (or would have been) similarly anomalous as units within an academic college, and they have obvious commonalities as university assets. Sharing the same reporting line to a University-level administrator would help ensure that their joint occupancy of the cultural district remained harmonious, productive, and aligned with the University's best interests.

The outcome of our April 30 meeting was that Nick Jones asked me to work with Erin to prepare a short white paper proposing a joint administrative structure for the arboretum and art museum. Nick expressed his belief that other units might benefit from having University-level administrative homes (such as other museums and the Center for the Performing Arts), and he speculated that the right solution might be an executive board to harmonize interests and relationships. Erin and I sent him a white paper at the end of July recommending that the arboretum and art museum jointly report to either an executive council, with a dotted-line report to the provost, or to a new vice president or vice provost for arts and culture. The white paper marked the end of my involvement in the future administration of the arboretum. I had pushed the question as far as I could without unduly meddling.

About a year later, in March 2021, Nick called me to a meeting to announce his decision to move the arboretum to Outreach after my retirement. The new director would report to the vice president for outreach, Tracey Huston, who was also present at the meeting. Nick said he expected that the art museum director would report to Outreach as well, but he had to speak to the dean first. Nick's decision aligned closely with our white paper, and the idea of making Outreach the home had been suggested to me more than once when I queried the Arboretum Steering Committee on this topic a year earlier.[2] Unfortunately, it later transpired that the art museum would not move to Outreach, and the Palmer Museum of Art continues to be administered by the College of Arts and Architecture. Tracey retired a year later, but her leadership during that year gave me confidence that Outreach would be a good home for the arboretum.

I retired on July 1, 2021, from my faculty position in the Department of Ecosystem Science and Management, concluding my directorship of the arboretum and ending what I have jokingly referred to as the longest service assignment ever for a Penn State faculty member—forty-six years, if counted from my appointment to the Penn State Arboretum Committee by Provost Larson. With agreement from Tracey Huston and Dean Rick Roush, Provost Jones appointed a faculty colleague, Sanford Smith, as interim director until a permanent director was chosen. Sandy was an excellent choice, and I believe he enjoyed this opportunity to participate in the arboretum project.

Tracey's office organized the search for a new permanent director with the assistance of a professional search firm. They selected Dr. Casey Sclar, who began the job on August 1, 2022, with a collateral appointment as an affiliate professor in the Department of Entomology. I had known Casey for many years as the executive director of the American Public Gardens Association, and his appointment assured me that the future of the arboretum would be in safe and capable hands.

As it happened, there were many other personnel changes around the time of my retirement, and the staff I was so proud to work with to build the arboretum have largely moved on. Of those who served for several years before I left, only Patrick Williams, Yuhui "Millie" Zhou, and Rebecca Horwitt remain. Alisha Barton left University employment in December 2019, and Linda Duerr retired in June 2020. Kelly Oleynik and Shari Edelson left near the end of 2022, Kelly to take another position within the University and Shari to pursue a PhD, as mentioned earlier. And Kate Reeder, who had been my first employee and partner since 2000, entered a well-deserved retirement in 2023. I have fond memories of all these colleagues and those who left or retired earlier—Cathy Kennedy, Andrew Gapinski, Nancy Stoner, Catherine Gowan, and others—all of whom labored enthusiastically to create The Arboretum at Penn State. It was immensely gratifying to work alongside such a fine group of people.

Roughly contemporaneous with a large turnover in personnel were other significant disruptions to the continuity of operations: disorder caused by Pollinator and Bird Garden construction from 2019 to 2021, extended closure due to COVID-19 in 2020, changes in administrative procedures following the move to Outreach in 2021, and access restrictions for two years (2021–23) during the construction of the art museum. However, the arboretum's large annual events have been resurrected, public educational programs are stronger than ever, fundraising has resumed its pace, and work is underway on a new master plan for the entire arboretum and a new garden, the first since 2021. The administrative transition has been successfully managed.

In a vision statement prepared by the oversight team for the 1999 Sasaki master plan, we wrote:

> In the 21st century, the Penn State Arboretum will . . . draw visitors, students, conference participants, visiting scholars, and new faculty . . . [and become] one of those special features of campus that strengthen personal attachments to Penn State. [The arboretum will benefit] programs in horticulture, forestry, biology, landscape architecture, engineering, ecology, entomology, geosciences, geography, and other environmentally oriented departments and colleges [and serve as] a convenient resource for a wide variety of educational, cultural, and social activities and events. Most importantly, the arboretum that we envision will "fit" Penn State—it will match, complement, and preserve the qualities and ambiance that make this campus unique and attractive to students, staff, faculty, alumni, and visitors.

We never revisited that statement to check our progress. Rereading it for the first time in over twenty years, I am a little surprised that our fancied description was actually prescient of what the arboretum has become. It is used annually by thousands of students and adults in university courses, public school classes, and outreach programs. It serves annually as a resource for usually a couple of dozen research projects in a variety of disciplines. It is a regular destination for field trips, tours, and social events as part of national and international conferences organized by university faculty. It is one of the top-ranked visitor destinations in the Centre Region, recognized as such by the local visitors' bureau, and consistently ranked #1 or #2 among "things to do" in State College by the popular website Tripadvisor ("Penn State" is sometimes in the first position). It is frequently named "their favorite place on campus" by students, employees, and community members, many of whom have commented to me that the arboretum is the best addition to the University and community in many decades.

I believed in our vision statement, but expressions of enthusiasm for the arboretum still surprise me a little. They are very gratifying, of course, but my main satisfaction in having completed the project comes from something else. Very soon after the Phase I construction fence came down, Susie and I walked around the botanic gardens and quietly enjoyed the view of the University campus from one of the benches facing the Fountain. This was on a Sunday evening in August 2009. The public had not yet discovered the gardens, and we were the only visitors. The occasion was our private celebration of the end of a project that had continually occupied my time and attention for fifteen years. Although I was raised and educated elsewhere, Penn State and State College hold my deepest affections for having afforded so many opportunities for a happy and rewarding career in such a wonderful community to live and raise a family. The greatest of those rewards has been helping create the arboretum and giving back, in some measure, to the University and community.

Appendixes

APPENDIX A

Timeline for the Creation of a Penn State Arboretum, 1855–1980

This timeline was compiled in 1997 by Lee Stout, University Archivist, Penn State University Libraries.

1855 General James Irvin donates two hundred acres and offers the option to buy an adjacent two hundred acres at a favorable price to trustees of the Farmers' High School of Pennsylvania.

1856–59 Professor of horticulture William G. Waring plants a variety of species of trees around Old Main and the core of today's campus, including many of the large American elms.

1860s–80s Tree planting continues, especially of American elms, Norway spruce, and Norway and silver maples. In the 1870s and 1880s, this is largely the work of Professor William Buckhout, who, in 1882, surveys plantings and reports 110 species and varieties on campus.

1867 Professor Francis Fowler and President John Fraser attempt to secure the services of Frederick Law Olmsted to design the campus of the Agricultural College of Pennsylvania. The college's impoverished state prevents it from hiring Olmsted.

1870s Professor of agriculture John Hamilton succeeds in leveling up much of the campus and planting the front lawn of Old Main with grass (it had been planted with potatoes and corn at various times). Hamilton suggests the hiring of landscape architect William Saunders to lay out the campus, but again, funds are lacking. Hamilton reports: "We were left to the mercy of amateurs and worse for locating our buildings, constructing our drives, and planting our trees and shrubbery. The result has been the cutting up of what were the most spacious grounds of any college in America into a few sections of common, rectangular city blocks."

1880s–90s Professors George C. Butz and J. Plummer Pillsbury continue tree planting around the campus. In 1911 Pillsbury surveys plantings and estimates 400 species and varieties on campus at that time. First major "building boom" on campus takes place in 1887–93, with a second building campaign in the 1900–1907 period. These take the college beyond one central building with a few barns, residences, and outbuildings, which were the entirety of the physical plant from the 1850s to the mid-1880s.

1907 First campus development plan accepted by the Board of Trustees. It is the creation of Charles Lowrie, a New York City landscape architect, and is the first plan to suggest a system of roads, walks, and building groups in a landscaped environment. Specifically, the plan proposes maintaining the large stands of trees on campus around the current Nittany Lion Inn and in Hort Woods, which was to be a park. To the east of Hort Woods would be a botanical garden and greenhouses, adjacent to future agricultural buildings north of Patterson Building.

1913 Board of Trustees selects the Philadelphia firm of Day & Klauder as college architects. Charles

Z. Klauder creates first campus plan. Klauder continues the initiative of Lowrie by adopting a Beaux-Arts campus design plan of highly geometric building placements, quadrangles, and roads. Klauder continues as designer of many campus buildings until his death in 1938.

1914 At its January 27 meeting, the Board of Trustees accedes to the requests of the Forestry and Landscape Gardening Departments and sets aside twenty-five acres for an arboretum in what is now the northeast corner of the golf course. No actual development takes place, however.

1924 Board of Trustees adopts a comprehensive plan for landscape development designed by Warren H. Manning of Cambridge, Massachusetts. Manning's plan includes (a) a scheme for central campus, with the principal object of beauty and fit with the educational values of the college as much as possible, and (b) "an arboretum to be located as may be suggested in the proposed comprehensive plan" and a nursery to propagate plant materials. Manning confides to a trustee, "An arboretum of the type of the Arnold Arboretum would not be practicable unless a very large endowment could be secured. The cost of maintaining the arboretum alone [would be] about $100,000 a year." The board requests that he make a further study of roads and paths, parking and other transportation issues, and future building expansion and plantings needed. Manning's plan initially calls for an arboretum to be located on the old golf course location. After much debate, he writes President John Thomas in October, saying, "I do not consider an arboretum advisable or practical on a general section set apart for this purpose. We can provide in general plantation [around the campus] all varieties of proven value and some experimental material."

1925 Faculty are critical of giving up the arboretum space on the golf course and the separate arboretum concept, but the administration follows Manning's advice.

1927 Professor of landscape architecture J. R. Bracken proposes a memorial arboretum and recreation area for Thompson Spring (on both sides of the current College Avenue between University Drive and Porter Road). The senior classes of 1927 to 1931 pledge their class gifts for the project, but because of the Depression, the college cannot collect enough funds to complete the project.

1930 Thomas W. Sears creates a landscape plan to complement Klauder's work, and it becomes the basis for plantings under the supervision of Walter Trainer in the 1930s and 1940s. Trainer reports that the formal geometric arrangement of buildings by Klauder is softened in Sears's plan to create a more informal appearance for the campus, with informal tree and shrub groups planted within the quadrangles. In 1934, Trainer reports that his survey shows 439 species of trees and shrubs. He begins the careful recording of the campus trees that results in his publication of *Campus Trees and Shrubs* (1946), with extensive maps. This clearly continues Manning's concept of the campus as arboretum. A botanical garden plan created by John Bracken for the Thompson Spring tract in the mid-1920s is still discussed, but no one can see any likelihood that it will reach fruition.

1938 Trustees renew approval of Bracken's Thompson Spring plan, but funding is still unavailable.

1972 A committee to create a comprehensive proposal for an arboretum is created by Provost Russell Larson, former dean of Agriculture.

1973 The committee's report says the campus, with more than 750 species, should be recognized as an "arboretum." In addition, a tract of about 320 acres (Farm No. 17 and part of Farm No. 13), stretching from north of the existing Henning Building to behind the Overlook Heights residential development, is identified as the most appropriate site for a dedicated arboretum. In September, the Board of Trustees reserves that space for the purpose.

1976 Continuing the committee's work, landscape architecture professor Glenn Steyers and his students finish a master plan for the arboretum in May in preparation for fundraising efforts.

1977 Long-range planning efforts begin to examine questions of whether the College of Agriculture or the Physical Plant should operate and maintain the arboretum and how much money would be necessary to operate it. Trustees authorize a fundraising effort of more than $9 million to construct the arboretum and its buildings and to create an endowment to maintain it.

1978–80 Fundraising activities move slowly, and an arboretum fundraising booklet is created in 1980 to move the effort forward. Unfortunately, President John Oswald loses interest in the project, and it begins to languish.

APPENDIX B

Historical Ownership and Land Use of the Arboretum Property

Early History

It is unlikely that the arboretum land was ever occupied on a long-term basis before European settlement.[1] The nearest source of open water was a small stream over one mile away (Thompson Run), which would have made daily living needlessly difficult, given the presence of other locations in the general vicinity that would have been more suitable for semipermanent villages and the cultivation of food crops. However, the area was surely visited by hunting parties, and it may have been used for that purpose by a succession of cultures over thousands of years.

Archaeological evidence indicates the emergence of a "Shenk's Ferry" culture within central Pennsylvania during the Late Woodland period (after about 1000 CE), although it is unclear whether they ever lived near what is now State College.[2] The Shenk's Ferry culture disappeared around the year 1600 when those people were conquered and perhaps absorbed by the Susquehannock, a more northern tribe. Susquehannock control of central Pennsylvania was brief, owing to frequent conflict with other tribes, and by the latter part of the seventeenth century most of the Susquehannock people had been absorbed by the Iroquois (Haudenosaunee). The Iroquois, along with the Lenape and other client tribes, controlled the region until European settlement. The arboretum property entered the colonial legal system with the 1754 Albany Purchase from the Iroquois and its reduction in 1758 by an amended treaty.[3]

The Hartley Warrant and the "Mitchell Tract"

During the eighteenth century, publicly owned land in Pennsylvania was transferred to private ownership through the purchase of a warrant for the survey of a tract of specified size. The tract became patented (titled) to the purchaser after the survey was filed. Although warrants were nominally limited to 400 acres, with any given individual limited to one claim, large blocks of contiguous tracts were sometimes amassed by a single owner who had arranged for surrogates to take out the warrants. One of the first land warrants in the State College area was for a 438-acre tract purchased by James Hartley in 1784 and surveyed on September 8, 1785. The northeastern end of the Hartley warrant became the Mitchell Tract portion of The Arboretum at Penn State (fig. B1).

Survey notes for the Hartley warrant indicate vacant land all around, so it appears that Hartley had free rein in laying out his "400-acre" homestead. We will never know why Hartley chose this tract and laid it out with eight sides. We do know that one side of the property lay on or near what was then the dividing line between Bedford County and Northumberland County to the north. Presumably, Hartley and his surveyor would have chosen to enclose any unforested and arable land (to minimize clearing for farming) as well as forested acreage for fuel and building materials. All eight traverse corners had trees that probably were selected for their prominence and promise of remaining life but also with an eye toward meeting the targeted 400 acres.

FIG. B1 The outline of the arboretum (center) superimposed on a map showing the Hartley warrant and fifty-three surrounding warrants, most or all acquired by Centre Furnace. Created by Samuel Grinstead based upon the *Original Warrant Index Map, Western Centre County* drawn by Edward J. Heary in 1996.

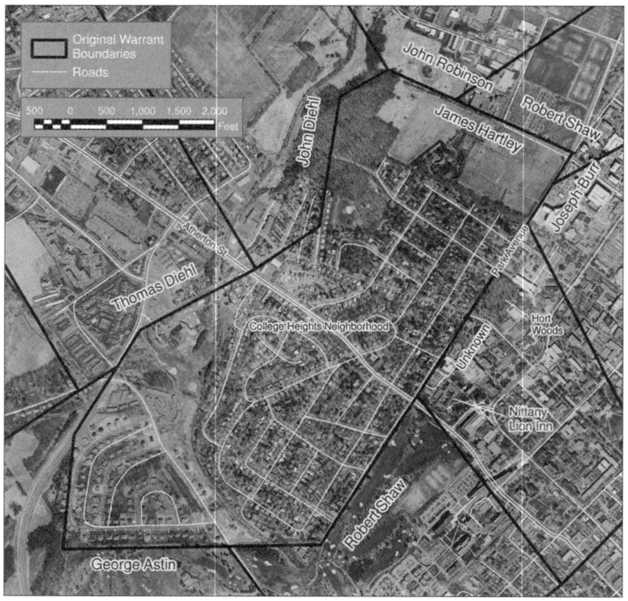

FIG. B2 Boundary of the Hartley warrant and surrounding Centre Furnace warrants overlain on a present-day aerial photograph with streets highlighted.

However they were chosen, Hartley's boundaries left their mark on State College's road system and the orientation of campus (fig. B2). The entirety of the College Heights neighborhood lies within the Hartley warrant. Its southeastern boundary became East and West Park Avenue, its southern and western edges coincide with the back of properties bordering Teaberry Lane and Ridge Master Drive, and its north-northeast trending boundary follows a section of Blue Course Drive.

Ownership of the Hartley property changed in 1837, when James Hartley sold the property to Frederick Krumrine.[4] Krumrine subdivided the land, and a 164-acre parcel containing part of the future arboretum was sold to John Neidigh in 1838. This land passed through several owners over the course of the rest of the century and was again subdivided by about half. In 1907, a tract of 89 acres was sold to H. Walton Mitchell, a Penn State alumnus who was then serving on the college's Board of Trustees. Thirteen acres of the Mitchell property were conveyed to State College Borough in 1982 to become part of Sunset Park. The remaining 76-acre tract was sold by Mitchell's granddaughters to the University in 1989, and this became the front part of the arboretum.

Centre Furnace and Later Ownerships

As shown in figure B1, the Hartley tract became surrounded by land warrants taken out by others, at least some of whom appear to have been acting as surrogates for speculators. Centre Furnace acquired most of these lands in the years 1790 to 1792, and in 1833 it purchased an additional large tract that included all parts of the future arboretum north of the Hartley property (the "Nicholas Diehl" tier of warrants in fig. B1). The furnace processed locally mined iron ore from 1792 to 1809 and from 1826 until 1858. Charcoal made from timber was used to fuel the furnace, and it is likely that all Centre Furnace lands were cut over for charcoal production at least once during the furnace's nearly fifty years of operation.

As business began to decline for Centre Furnace, its owners (James Irvin and later Moses Thompson, his brother-in-law) began to subdivide their unused lands as "farms" and offer them for sale. In 1855, Irvin offered to donate 200 acres of land for the creation of a Farmers' High School of Pennsylvania. The offer was accepted, along with an offer to sell an additional 200 acres, and both parcels were deeded to the school in 1857. Together, those 400 acres became the core of the modern Penn State campus just south of Park Avenue and adjacent to the old Hartley warrant.[5] By 1859, when the school opened its doors to students (and a year after Centre Furnace closed), Thompson had acquired full ownership of the furnace lands and completed their conversion into farm tracts, all of which were eventually sold off.

In addition to the Mitchell Tract, The Arboretum at Penn State sits on portions of four Centre Furnace farm tracts purchased by Penn State between 1922 and 1946 (fig. B3). The boundaries of the Mitchell Tract and those four nineteenth-century properties are still traceable in twenty-first-century landscape features (fig. B4). The oldest distinctive relic of past land use is probably the fencerow that separates the former Rockey and Foster properties. This dates at least to 1851, when what we know as the Rockey Farm was sold from Centre Furnace holdings to William Blair as a farm.

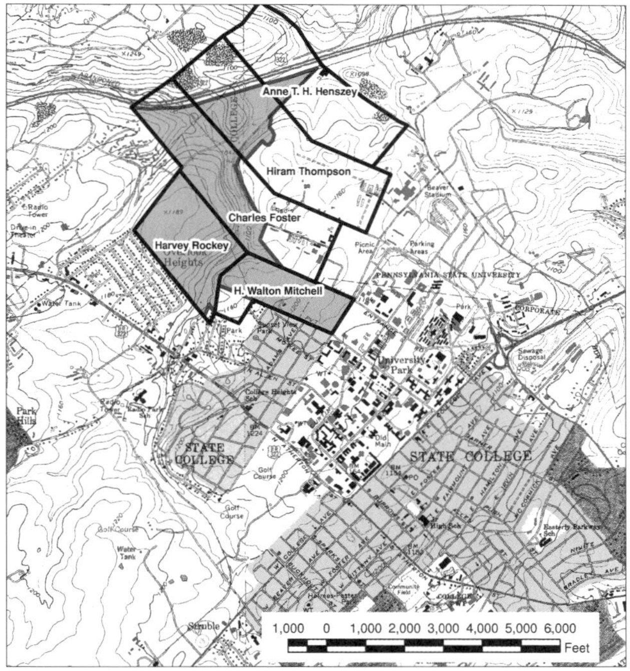

FIG. B3 The five properties constituting the arboretum (shaded area) as they were when acquired by Penn State during the twentieth century: Thompson (Penn State Farm 8, 231 acres, acquired in 1922), Foster (Farm 13, 204 acres, 1935), Henszey (Farm 16, 175 acres, 1935), Rockey (Farm 17, 128 acres, 1946), and Mitchell (Farm 32, 76 acres, 1989).

The Bellefonte Central Railroad

The arboretum is traversed by about a mile of the grade for a former railroad (now a pedestrian/bicycle path) that was built in part to connect Bellefonte with the community that became State College a few years later. The railroad began carrying passengers in 1887 with a stop at "Krumrine Station" located along what is now North Atherton Street at the base of the hill just north of campus. The railroad was extended around the hill to a station at the edge of campus along College Avenue in 1892. The Bellefonte Central Railroad carried passenger traffic to State College until 1946 and freight until 1974, after which the right of way was abandoned. The rails were removed in 1976, leaving a corridor that was used informally for the next thirty years by hikers and cyclists as vegetation gradually encroached on the old railroad bed.[6]

Hartley Wood, Other Wooded Areas, and Past Land Use

All 370 acres of the arboretum had been logged, cultivated, or pastured during a century or more of use before the land was acquired by Penn State, and the land has been further changed over the thirty to one hundred years that the University has owned it. Penn State uses the upper portion of the Rockey Farm (fig. B3) to grow crops, probably continuing earlier practices dating back at least to the sale of the land by Centre Furnace in 1851. The limestone-derived soil there is identical to that underlying the H. O. Smith Botanic Gardens on the Mitchell property and typical of the richer farmland in this area. Except for the area around the Schreyer House (called Lisnaward when the Mitchells owned it), the nonwooded portion of the Mitchell Tract was variously farmed and pastured in the past, and the University leased it for many decades from the Mitchell family for that purpose.

Unlike the Rockey Farm, the soil on most of the Foster Farm within the arboretum is poorly suited for growing crops. The University pastures horses on its northern end, near the highway, and an American chestnut breeding orchard is located southeast of the pasture. Other portions of the Foster Farm that were formerly pastured and

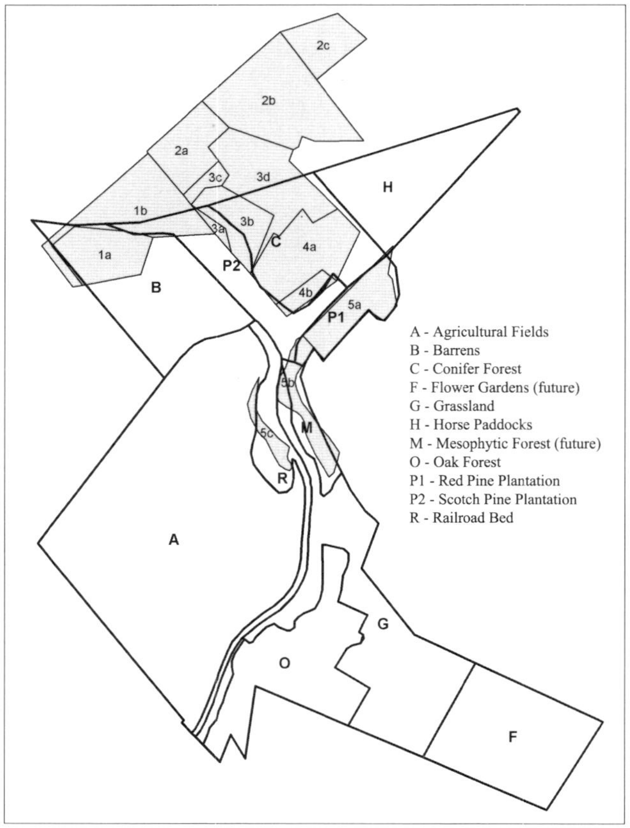

FIG. B4 Residual evidence of nineteenth-century property boundaries in and around The Arboretum at Penn State.

FIG. B5 College Farm Woodlot compartments (in gray) overlain on vegetation zones (capital letters). Compartment descriptions, from information compiled in 1928 and 1938 and preserved in the University Archives, were as follows: 1a (old field), 1b (scrub oak with patches of aspen and pitch pine), 3a (scattered hardwoods underplanted with Norway spruce after 1928), 3b (experimental conifer plantations postdating 1928), 3d (red and eastern white pines planted in 1928), 4a (mixed stand of hardwoods and conifers, part of it cut over twenty to forty years before 1938 and in sprout clumps), 4b (eastern white pine planted in 1928), 5a (red and eastern white pines planted in 1927 and 1928), 5b (described in 1938 as containing thirty- to forty-year-old hardwoods with scattered pitch pine and eastern white pine), 5c (described in 1938 as containing sixty- to eighty-year-old white oak and scarlet oak in degraded condition due to fires and cattle grazing). Compiled by Samuel Grinstead from maps drawn in 1938 and overlain on vegetation zones in Daniel Laughlin, *Floristic Inventory of The Arboretum at Penn State*, Special Publication No. 1 (University Park: The Arboretum at Penn State, 2001). Compartment descriptions and maps are from D. D. Stevenson, H. A. Meyer, and L. J. Bennett, *Management Plan for College Farm Woodlands* (n.p., 1939). Descriptions were supplemented with information on a 1928 map drawn by faculty in the Department of Forestry.

abandoned have been largely taken over by invasive shrubs. This is also true of the steep portion of the Rockey Farm close to the bottom of Big Hollow. The arboretum's prairie re-creation project site on the south side of Big Hollow is part of the former Foster Farm and was also formerly pastured. It, too, was densely covered with shrubs before restoration began.

The arboretum has only four significant, naturally wooded areas, not counting fencerows, tree plantations, and scattered trees. All four of these areas are dominated by two or three species of oak along with scattered eastern white pines. Black cherry and hickories are also common in the overstory. Two of these stands are on the Foster Farm and stretch linearly along either side of Big Hollow (5b and 5c in fig. B5). Their trees date to the decades between 1860 and 1910, according to fieldwork and mapping performed in 1938.[7] Stand 5b in figure B5 appears in aerial view to merge with 5a to make a single wooded stand along the southeastern edge of Big Hollow Road, but the two stands are different: 5b (on the Foster Farm) consists of naturally regenerated hardwoods, as described, and 5a (on the Thompson Farm) is a conifer plantation.

The third *naturally* wooded area is a 9-acre stand of native hardwoods and eastern white pine within the Thompson Farm property (4a in fig. B5). This area had been partially cut some twenty to forty years before it was mapped in 1938, and it is now a mature forest stand. In the 1970s and 1980s, when I took classes to this area on field trips, it was still possible to find stunted and struggling saplings of American chestnut in the understory. Chestnut blight (which usually infects only larger trees) made that an uncommon occurrence then, and it is even rarer now.

Except for an open area adjoining Big Hollow Road (fig. B5), the remainder of the surrounding Thompson Farm property was planted with various conifers in 1927 and 1928 or shortly thereafter. These plantations visually merge with the stand described just above to form a continuous forest tract to the northwest of Big Hollow Road and in prominent view from the State College Bypass. The planted trees (mostly eastern white pine, red pine, and Norway spruce) are now quite large. Some have died, and hardwoods are invading, but a vestige of rows and regular spacing reveals their origin as plantations. They were installed by faculty and students in the Department of Forestry, which was using the site for research

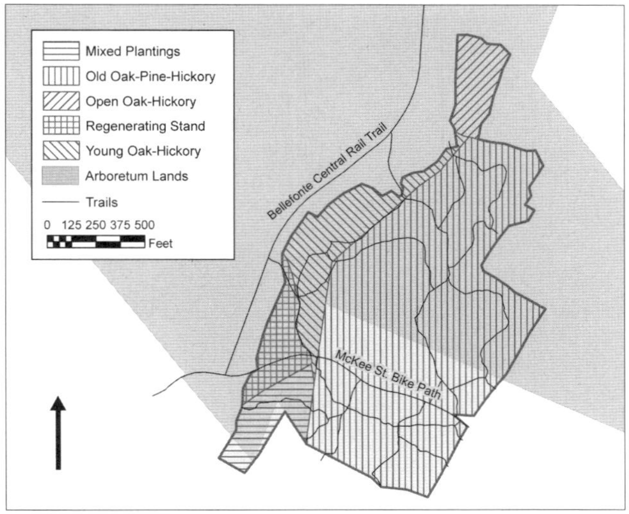

FIG. B6 Stand composition of Hartley Wood in 2007 based upon differences in age and species composition. The unshaded lower portion is a portion of the Mitchell property deeded to State College Borough in 1982 and now part of Sunset Park. Hartley Wood is visible near the top of the aerial photograph in this appendix.

and teaching. The land they are on must have been free of trees at the time of its purchase in 1922, and it was probably used by the previous owners as pasture. This forested hillside has the same non-calcareous, sandstone-derived loam as on the Foster Farm and would have been poorly suited for farming. The open area along Big Hollow Road is on a colluvial flat where there was a farmhouse and plowed field in 1938.

The fourth naturally wooded area is Hartley Wood on the Mitchell Tract, named after James Hartley. As a tract of continuous forest studied by Samuel Grinstead, the woodlot comprises five stands distinguished by age and species composition (fig. B6). They also have different ownership histories, and some were never owned by Hartley.

The narrow, 2-acre stand on the north side, labeled "Open Oak-Hickory" in figure B6, is on land formerly owned by Centre Furnace and purchased by Penn State as the Foster Farm. The oldest of its trees date to about 1860. They now provide scattered shade to the core of the arboretum's prairie re-creation project begun in 2010, and the arboretum does not count this as part of Hartley Wood. The three stands bordering the west side (totaling 12 acres) are also on former Centre Furnace land and came to the University as the Rockey Farm. These three stands originated in the twentieth century, and the "Regenerating Stand" was an open field when I first visited this site in the fall of 1974.

The core of Hartley Wood is the 32-acre "Old Oak-Pine-Hickory" stand in figure B6, and an 18-acre portion of this stand was part of the Mitchell Tract purchase in 1989. The 13-acre balance is part of State College's Sunset Park, but both parcels are identical in stand composition and were part of the Hartley warrant with the same ownership history until 1982.[8]

The core stand contains old trees scattered across both ownerships. There are six trees that predate the 1784 Hartley warrant. Three of those trees are estimated to have germinated in the last half of the seventeenth century (the earliest in 1661).[9] Another fourteen trees are younger but predate 1835, making a total of twenty trees (all white oaks, *Quercus alba*) that are at least 190 years old as this is written. From these facts we know that Hartley Wood was never completely harvested. It is a rare example of a central Pennsylvania valley-floor forest containing trees that were living at the time of the earliest European settlement, and the arboretum's 18-acre portion is its most significant natural area.

We do not know how Hartley used his property—whether he farmed it or even lived there at all, and whether he sold timber from parts of the property to Centre Furnace for charcoal-making. It is impossible to know whether Hartley harvested *any* trees from our woodlot, but we can say with high confidence that partial cuts were made around the time of the 1837 and 1838 sales and afterward, because a very high proportion of the present forest canopy originated at that time, indicating the occurrence of one or more partial harvests (or other disturbances) that created growing space for a cohort of new trees. We are fortunate that remnants of the primeval forest have survived, though, and even the canopy's "young" cohort of oaks, hickories, and eastern white pine is approaching the end of its second century. Whether Hartley Wood survived by design or neglect we do not know, but its very existence is attributable to James Hartley's ownership. If Hartley had not arrived there first, Centre Furnace would have purchased the land and harvested its forest for charcoal production. At best, the woodlot might have survived as something resembling Hort Woods on campus, on land once owned by the Centre Furnace. Hort Woods is nice, but not *ancient*.

APPENDIX C

Arboretum Master Plan by Sasaki Associates (1999)

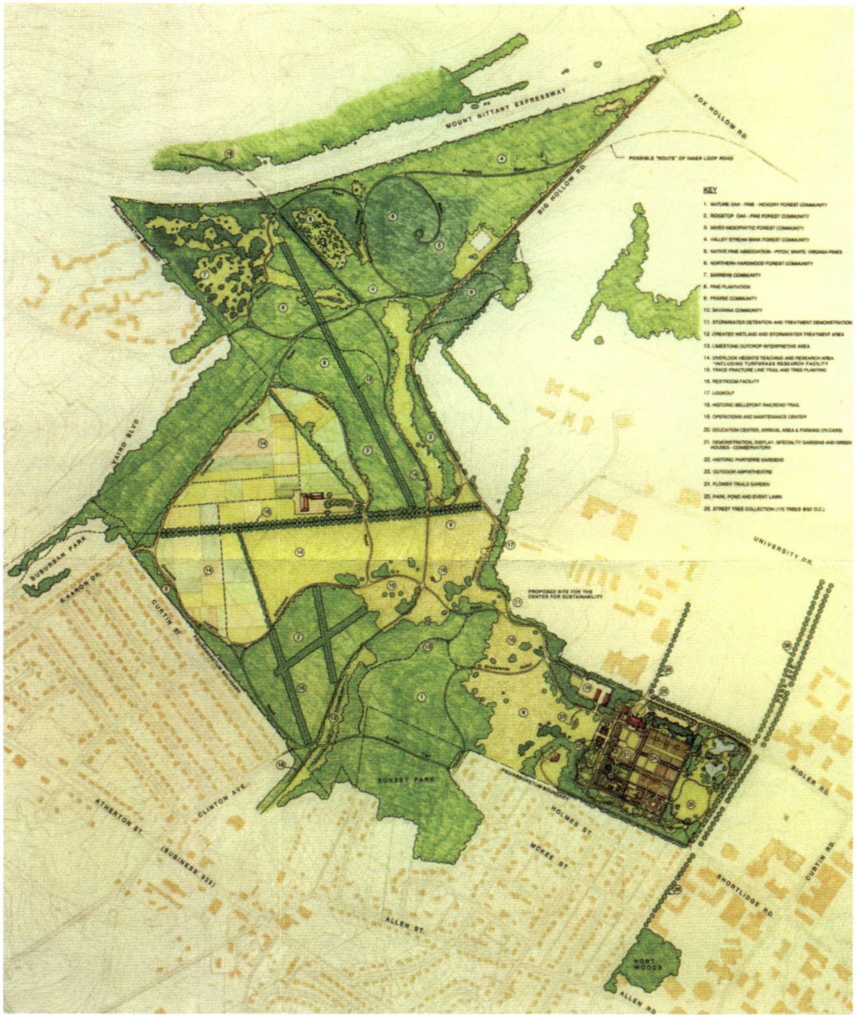

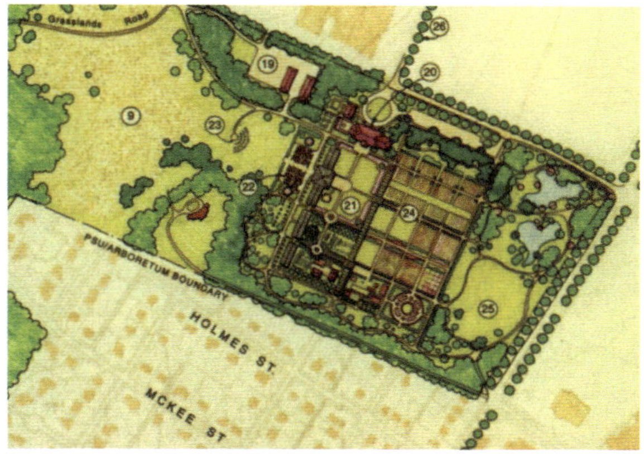

LEGEND

9 Prairie Community
19 Operations and Maintenance Center
20 Education Center, Arrival Area, and Parking (75 Cars)
21 Demonstration, Display, Specialty Gardens and Greenhouses/ Conservatory
22 Historic Parterre Gardens
23 Outdoor Amphitheatre
24 Flower Trials Garden
25 Park, Pond, and Event Lawn

APPENDIX D

Mitchell Tract Program Summary

This summary was prepared by the author in September 2000 as background for a master plan.

The Mitchell Tract and the Arboretum Master Plan

The Mitchell Tract property is fundamentally important to proposed development for the Arboretum at Penn State. Although it occupies only 18 percent of the arboretum land, Mitchell Tract development represents 60 percent of the $34 million projected capital development program that will transform existing agricultural lands and woodlots into a first-class university arboretum.

The arboretum property comprises 395 acres of land bordering the Penn State campus, US 322 (soon to be Interstate 99), and two major residential neighborhoods. The property consists of two principal upland areas and a central, dry hollow that runs approximately north and south through the length of the arboretum. The Mitchell Tract makes up one of those upland areas. It grades slightly uphill from Park Avenue before descending about 115 feet into Big Hollow. The tract is currently in pasture grasses, with a 170-year-old woodlot bordering the western edge.

The Arboretum at Penn State will further the creative use of designs, materials, and practices that preserve the beauty and functions of nature in urbanizing landscapes, as stated more generally in its mission in the master plan. The arboretum itself is intended to become a successful case-study of its own mission because of the need to unify elements of the human-nature continuum either within the property or along its borders: residential neighborhoods, commercial development, a major transportation corridor, formal gardens, agricultural landscapes, forestry plantations, and natural plant communities. The design of the arboretum reflects Penn State's land grant heritage, and the mission of the arboretum will be accomplished through Penn State's land grant philosophy of improving the lives of people through teaching, research, and outreach. Thus, the arboretum is both a public cultural resource and an institutional, intercollege facility for research and education.

A Vision for the Mitchell Tract and Schreyer House

For most persons, the Mitchell Tract will be the heart of the arboretum, although those who visit only this area will regard it more as a botanical or horticultural garden than an arboretum. This area will be a place of particular beauty within the arboretum, and its design will be evocative of serenity, peace, and contentment. However, this area will also be the locus of aggressive teaching, outreach, and research programming, and the gardens that support these programs will evoke curiosity, interest, enthusiasm, and comprehension. Thus, the landscape of the Mitchell Tract will challenge both our emotions and our intellect.

As both the public face of the arboretum and its principal attraction for visitors, the Mitchell Tract landscape will reveal enough of the arboretum from Park Avenue to invite visitation, but it will also be intimate enough to encourage and reward repeated visits. Public opinion of the entire arboretum will be largely shaped by two aspects of just the Mitchell Tract landscape: (1) its overall appearance from Park Avenue and (2) the quality of the gardens, collections, and facilities within ten minutes' walk of the arboretum parking lot. For this reason, the Mitchell Tract landscape will be designed with extraordinary taste and a significant degree of restraint;

it will be immediately compelling and yet, through subtlety and complexity, reflective of the quality we wish to convey.

Because of its size, the relocated flower trials garden will be a significant, even perhaps controlling, feature of the Mitchell Tract landscape, but it will not be a dominant feature. The flower trials garden will be fully integrated with the design of other gardens and collections, so that the necessarily functional nature of the flower trials garden does not detract from an overall design goal that places a premium on beauty.

The constructed and formalized landscape of the eastern half of the Mitchell Tract will stand in contrast to the completely "natural" (if managed) prairie patch and woodlot landscapes of the western half and Big Hollow South. Thus, the human-nature contrast will be maximized in this, the most visited, portion of the arboretum. In recognition of our mission, the integration of the human landscape into the natural landscape within this portion of the arboretum will be sensitive and deliberate, and this transition will be one of the most surprising and remarkable features of the arboretum.

The Mitchell Tract gardens and facilities will likely be the principal source of revenue generation for the arboretum, and the landscape plan will be fully cognizant and supportive of this role. Likely sources of revenue are rental or staff use of the education center for conferences, short courses, and evening classes, and rental of the gardens and outdoor facilities for weddings, receptions, and similar events.

Whether for revenue generation or not, the design and composition of the gardens will encourage the widest possible use of the arboretum by Penn State's various academic programs. The following academic departments may have active or latent interest in the Mitchell Tract plan and should be consulted: agricultural and biological engineering, agronomy, animal science, architectural engineering, architecture, biology, civil and environmental engineering, entomology, forest resources, geography, geosciences, history, horticulture, landscape architecture, medieval studies, plant pathology, visual arts, and women's studies. Non-departmental units that should be consulted include the Biotechnology Institute, Center for Biodiversity Research, Center for Environmental Chemistry and Geochemistry, Center for Medieval Studies, Center for the Performing Arts, Center for Sustainability, Center for Watershed Stewardship, Earth Resources Research Institute, Hamer Center for Community Design Assistance, Life Sciences Consortium, Max Kade German-American Research Institute, and Pennsylvania Transportation Institute. We will facilitate this consultation.

The landscaping surrounding the adjacent Schreyer House, which serves as the university president's residence, will be fully integrated with that of the arboretum. The arboretum will be accessible to Schreyer House visitors, but visitors to the arboretum will sense that the Schreyer House property is off-limits to the general public. The master plan for the Schreyer House shall take into consideration both the public and private needs of the president.

The Plan for the Mitchell Tract

The plan for the Mitchell Tract property will refine and perhaps enhance the conceptual design for this area presented in the master plan developed by Sasaki Associates in 1999. It is expected that the consultant will work with the arboretum director and a small advisory group of Penn State faculty and staff to develop a plan that considers six categories of elements and their particular expression in this part of the Arboretum at Penn State: site attributes, visitor experience, collections and gardens, structures, programming, and infrastructure. Our preliminary judgments and recommendations are summarized below.

I. Site Attributes

Context

- The Mitchell Tract is the most visible and public portion of the arboretum; the quality of design and its execution will be the major determinants of the perceived quality of the arboretum.
- As the principal public entrance to the arboretum, the Mitchell Tract gardens and collections must not only be attractive and captivating, but their design must encourage visitors to depart this area and explore the more remote and ecologically natural portions of the arboretum.
- This portion of the campus is dominated by open distances, so intimacy will be an important attribute to develop and emphasize within this portion of the arboretum.
- The Schreyer House "peninsula" into the Mitchell Tract property creates the possibility that visitors will see the arboretum as much smaller than it really is, and this effect must be mitigated to the extent possible.
- It is anticipated that funding for construction will not come from a single, major donation, so the design must

permit construction in segments that must visually "occupy" the area as quickly as possible and then fill in as funding materializes.
- Pedestrian access to the arboretum from campus must be encouraged and facilitated.

Views and Sight Lines
- Buildings along Services and Big Hollow Roads, especially the side and rear of Housing and Food Services Building, are significant negative visual elements that must be screened or managed in some way.
- Park Avenue is slightly elevated above adjacent portions of the Mitchell Tract, affording a good view of the property up to approximately the crest in the vicinity of Services Road. Views to and from the buildings and parking garage planned for the other side of Park Avenue will become important as that area is developed.
- The view from the crest down the long slope into Big Hollow (at least to the woodlot) is important.

II. Visitor Experience
- Visitors should be permitted to experience the gardens on their own and in an unprogrammed, individualized fashion.
- Visitors must be encouraged to venture away from the formal gardens; the distance from the crest of the Mitchell Tract to the woodlot must be made to seem shorter than it is.
- It is altogether realistic to strive for a facility that will be one of the most desirable leisure-time destinations in central Pennsylvania.

III. Collections and Gardens
- Our goal is to collect and display the best plants for landscaping use in central Pennsylvania, with a special emphasis on native plants and plant germplasm.
- Specialty gardens and collections bulleted in the master plan should be regarded as a preliminary list that must be expanded and refined through further discussion with faculty and constituency groups.
- Gardens and living collections should reflect strengths and uniqueness of Penn State and our Pennsylvania heritage: *e.g.*, agriculture, horticulture, forestry, turf grass science, landscape architecture, William Penn's colony, our unique ethnic heritage, etc.
- Gardens and living collections must support teaching, outreach (extension), and research programming in appropriate departments.
- The plan must include a strategy for adding gardens (infilling) as funding becomes available.

IV. Programming
- Education is key to the mission of the arboretum and its role as a university facility, and it must be encouraged and facilitated by the design of the arboretum. Education, both residential teaching and extension/outreach, must be *planned for* in the design of the Mitchell Tract.
- We should think creatively about educational opportunities not normally associated with gardens (*e.g.*, soil profile display for public education and courses in Soils and Geosciences).
- Programming will include environmental impacts of ornamental landscaping, including considerations of invasive exotics and soils and water contamination.
- It is expected that the Mitchell Tract gardens, collections, and facilities will be used for a variety of cultural and social events, which may become sources of revenue to cover operating expenses. These opportunities need further definition.

V. Infrastructure
- Are projected sizes for the education center, operations and maintenance center, and parking appropriate?
- Plans for an amphitheater should be seriously examined with regard to function, access, and possible impact on the neighborhood.
- Outdoor lighting will be essential for parking and receptions in the lawn behind the education center. The need for other lighting must be considered, and consideration must be given to mitigating the impact of lighting on the neighborhood.
- Visitors must have access to water and restroom facilities.
- There may be an opportunity to move the operations and maintenance center to behind Housing and Food Services.
- All facilities will be conceived, designed, and operated with state-of-the-art attention to stormwater management; water, soil, and nutrient management; landscape maintenance; energy conservation; and other issues of sustainability.

APPENDIX E

First Master Plan for the Mitchell Tract by Marshall·Tyler·Rausch (2002)

APPENDIX F

Design Principles, Design Guidelines, and Standing Review Committees for The Arboretum at Penn State

April 18, 2005

This document was prepared at the request of Kim Steiner, Arboretum director, by an Arboretum Design Guidelines Task Force comprised of the following University faculty and staff members:

Robert Berghage, Associate Professor of Horticulture
Robert Cooper, Manager of Engineering Services
Peter Everett, Associate Professor of Marketing
Eliza Pennypacker, Professor of Landscape Architecture, Task Force Chair
James Sellmer, Assistant Professor of Ornamental Horticulture
Gordon Turow, Director of Campus Planning and Design
Scott Wing, Associate Professor of Architecture
David Zehngut, University Architect

"The mission of the Penn State Arboretum is to promote the quality of life by seeking, through scholarship, research, and education, collaborative solutions to growing demands on the natural landscape and its resources. As a place of beauty in a rapidly expanding metropolitan area, an educational facility, and a microcosm of the human/nature continuum in settled landscapes, the Penn State Arboretum shall strive to become an interdisciplinary 'institute for land health' of state, regional, and perhaps national significance."

—Excerpt, *Preliminary Master Plan for The Penn State Arboretum*, Sasaki Associates, Inc., January 1999; hereafter, *Sasaki Preliminary Arboretum Master Plan*

DESIGN PRINCIPLES[1]

I. The Arboretum will be an integral part of a major gateway to the University and a vital part of the fabric of the University Park campus, linked both physically and symbolically to the University and surrounding community. Its design shall feature open and porous edges, creating a welcoming environment and facilitating accessibility from both the campus and surrounding neighborhoods.

II. The architecture and landscape design of the Arboretum shall respond to the cultural traditions and natural forces rooted within central Pennsylvania's environment, its people, climate, and resources. Design elements shall recognize the Arboretum as a unique entity within the campus, while responding to both regional design character and local campus character (especially that of the contiguous East Subcampus and other development along Park Avenue).

III. The Arboretum will be contemporary and innovative through a harmonious composition of disparate elements along the human/nature continuum, from the most clearly human-controlled zone along Park Avenue to the most nature-controlled demonstrations of natural systems in the northern zone of the site. Throughout, the Arboretum will serve as a metaphor for environmentally compatible use of the natural landscape by humans. The design expression of the human/nature continuum shall be varied in accordance with the three zones established by the *Sasaki Preliminary Arboretum Master Plan*: the Mitchell Tract, Overlook Heights

Teaching and Research Area, and the Big Hollow, providing distinction to each zone.

IV. Designs for the Arboretum shall emphasize four interpretive themes: 1) Richness of biological and ecological diversity in native and cultivated collections; 2) Stewardship and conservation of soil, water, and biological resources; 3) Demonstration of environmentally responsible landscape design; and 4) Restoration of degraded landscapes.

V. All development within the Arboretum, including all research and teaching activities, shall be grounded in exemplary practices of stewardship, conservation, and, where appropriate, preservation.

VI. All elements of the Arboretum, including all research projects and facilities, shall be designed to enhance and contribute to the educational and outreach missions of the University and the Arboretum.

DESIGN GUIDELINES

I. Conservation and Stewardship

A. Designers shall obtain LEED (Leadership in Energy and Environmental Design) certification for major new building projects.

B. Design, construction, and maintenance of Arboretum facilities shall comply with best management practices for stormwater management and water quality in order to achieve the following objectives:

1. Protection and enhancement of groundwater recharge capacity;
2. Preservation of natural drainageways;
3. Management of stormwater as a site resource through innovations in on-site irrigation, maintenance activities, and groundwater recharge;
4. Compliance with wellhead protection area management initiatives.

C. Architects shall consider life-cycle costs in addition to initial construction costs.

D. All Arboretum planning shall support Penn State's policies regarding environmental stewardship (currently "Key Initiative 1: Environmental Stewardship" in *Finance & Business' 2002–2005 Strategic Plan*).

II. Education

A. Projects, buildings, infrastructure, gardens, and plant collections should include educational materials that explain both the interdisciplinary design process and the completed project for the student and the public.

B. Whenever possible, architects and landscape architects shall strive to involve students and the public in the design process of each new project in the Arboretum.

III. Design of Primary Landscape Spaces

A. Design of each landscape area of the Arboretum shall respond to the character prescribed for that locale in the *Sasaki Preliminary Arboretum Master Plan* and the *Master Plan for the Mitchell Tract in The Arboretum at Penn State* by Marshall·Tyler·Rausch (April 2002; hereafter, *MTR Mitchell Tract Master Plan*).

B. Selection of plant materials (e.g., ornamental vs. native) and arrangement of plant materials (e.g., geometric vs. naturalistic) shall be appropriate for the landscape character of that part of the Arboretum. Species known to be invasive in central Pennsylvania shall not be planted.

C. Architects and landscape architects shall establish strong compositional unity within each landscape space as well as visual continuity between contiguous landscape spaces.

IV. Site Design of Primary Building Elements

A. Architecture and landscape design in the Arboretum will respect the context of the area. This includes the aesthetics, scale, and quality of the buildings and civic spaces on campus, as well as the agricultural history and activities of the region.

B. Landscape elements and buildings, including structures such as parking areas, entry walkways, porches, and terraces, shall be organized and designed to spatially define, order, and sequence these elements and structures with respect to one another.

C. Building and landscape elements shall be designed and constructed with a view to their relationship with future elements as described in the *Sasaki Preliminary Arboretum Master Plan* and *MTR Mitchell Tract Master Plan*.

V. Architectural Form and Character; Relationship of Interior and Exterior Spaces

A. The design of architectural and landscape elements shall respect regional and cultural traditions while symbolizing the vision and innovative nature of the institution. Direct stylistic imitation of historical structures shall be avoided.

B. Architectural design shall emphasize the relationship between interior and exterior spaces and maximize the use of natural light.

C. The exterior scale, composition, color, and materials of buildings shall complement and enhance the surrounding landscape of the Arboretum.

D. Architects shall use transitional spaces such as porticos, pergolas, and sunrooms to create inviting connections between interior and exterior spaces.

E. Educational, research, and social venues shall accommodate audiences of various sizes simultaneously and without visual or audible competition between venues and audiences.

VI. Materials and Building Systems

A. Materials and building systems shall be durable (to minimize life-cycle costs) and adaptable (to accommodate future change and growth). Whenever possible, these materials will be authentic, reflecting their substance and structure in their appearance, rather than simulating another substance.

B. Materials in Arboretum structures shall adhere to the University's standards for construction quality and craftsmanship, as articulated by the Office of Physical Plant. These requirements can be accessed on the Internet at www.opp.psu.edu/stnd/stnd.htm.

VII. Circulation

A. The principal circulation system in the Arboretum shall be pedestrian and bicycle trails to maintain the integrity of the site and the visitor's experience of an "oasis of green and peace in a rapidly urbanizing landscape."

B. In general, the circulation system of the Arboretum shall fit existing continuous, curvilinear alignments that evoke a sense of harmony with the natural setting. The only exceptions should be the "fracture trace" trails (*Sasaki Preliminary Arboretum Master Plan*) and the existing straight-line sections of Big Hollow Road.

C. The design of the circulation system shall preserve the natural woodlands, natural landforms, native flora, and native fauna of the Arboretum.

D. Maintenance, service, delivery, and emergency vehicles shall be sensitively accommodated in the planning and design of the circulation system, preferably by adapting pedestrian and bicycle paths for their use, whenever feasible.

VIII. Signage, Lighting, and Outdoor Furnishings

To ensure visual unity and institutional clarity, signage, lighting, and outdoor furnishings (such as benches, trash receptacles, and bicycle racks) for any area of the Arboretum shall relate stylistically to existing University Park and Arboretum standards. Lighting should be designed to minimize light pollution of the night sky.

ADDITIONAL RECOMMENDATIONS

In addition to design principles and guidelines, the Arboretum Design Guidelines Task Force recommends that the following standing committees be established to provide oversight in the design/development review process to ensure the design integrity of the Arboretum:

I. Arboretum Design Committee

An Arboretum Design Committee shall be appointed by the Arboretum director to work with the Office of Physical Plant in developing the Arboretum. The committee shall review all development plans as "acting stewards" of the *Sasaki Preliminary Arboretum Master Plan*, the *MTR Mitchell Tract Master Plan* (if appropriate), and the Arboretum vision, mission, concept, principles, and guidelines.

II. Arboretum Horticultural Committee

An Arboretum Horticultural Committee shall be appointed by the Arboretum director to review proposed materials and locations of horticultural collections within the Arboretum, including research collections. This committee shall review all horticultural proposals as "acting stewards" of the *Sasaki Preliminary Arboretum Master Plan, the MTR Mitchell Tract Master Plan* (if appropriate), and the Arboretum vision, mission, concept, principles, and guidelines. The Arboretum Horticultural Committee shall serve as advisors to the Arboretum Design Committee and the Arboretum director.

III. Arboretum Research Committee

An Arboretum Research Committee shall be appointed by the Arboretum director to review and approve research projects proposed for locations within the Arboretum. This committee shall review all proposed research projects to ensure the compatibility of the research project intent/requirements and the Arboretum mission, especially in regard to education and environmental stewardship. The Arboretum Research Committee shall serve as advisors to the Arboretum Design Committee and the Arboretum director.

APPENDIX G

Mitchell Tract Master Plan Revision by M·T·R Landscape Architects (2006)

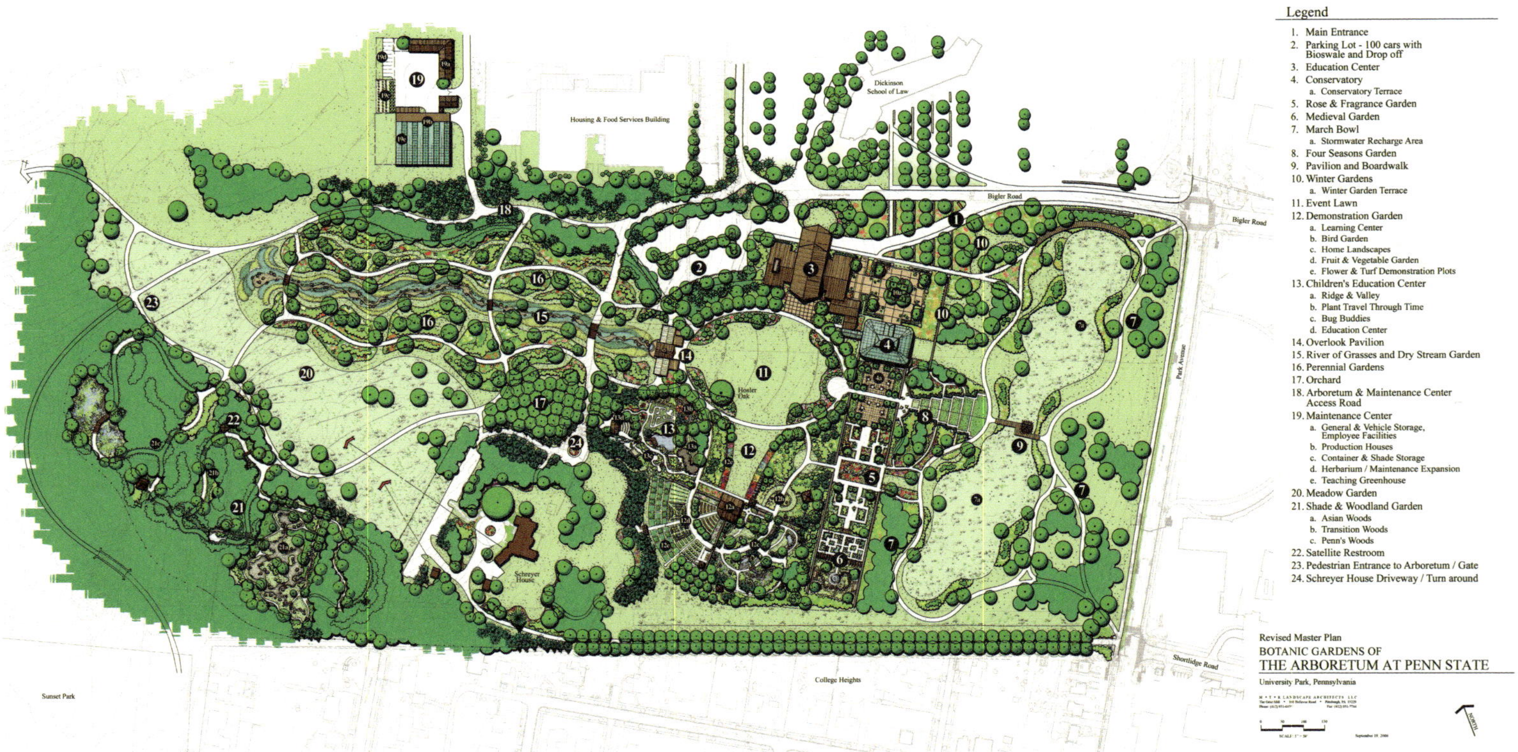

APPENDIX H

Planned Extent of the H. O. Smith Botanic Gardens (2006)

APPENDIX I

Final Version of the Mitchell Tract Master Plan Prior to Construction (2008)

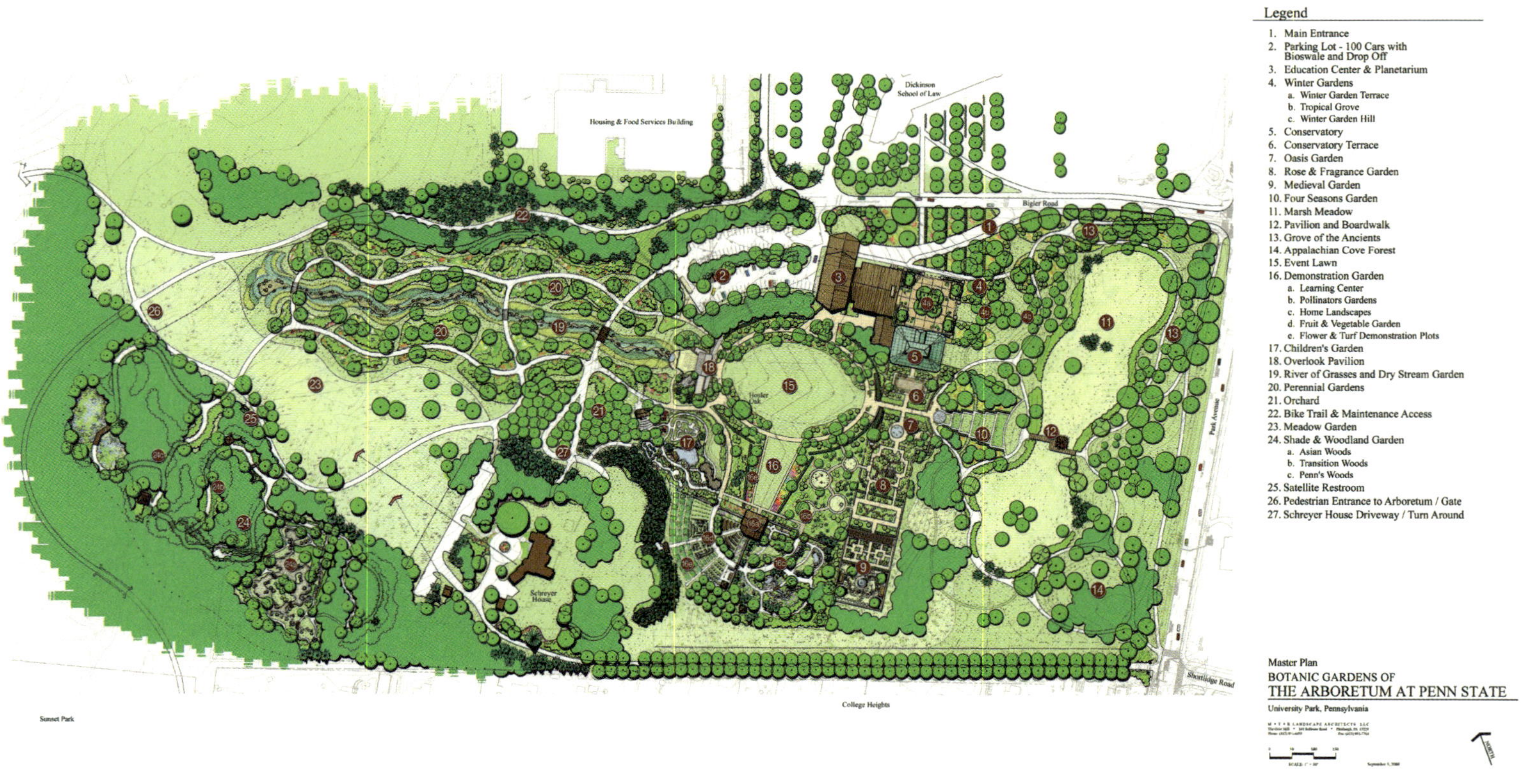

APPENDIX J

Map of the Marilyn Quigley Gerhold Wildflower Trail Complex

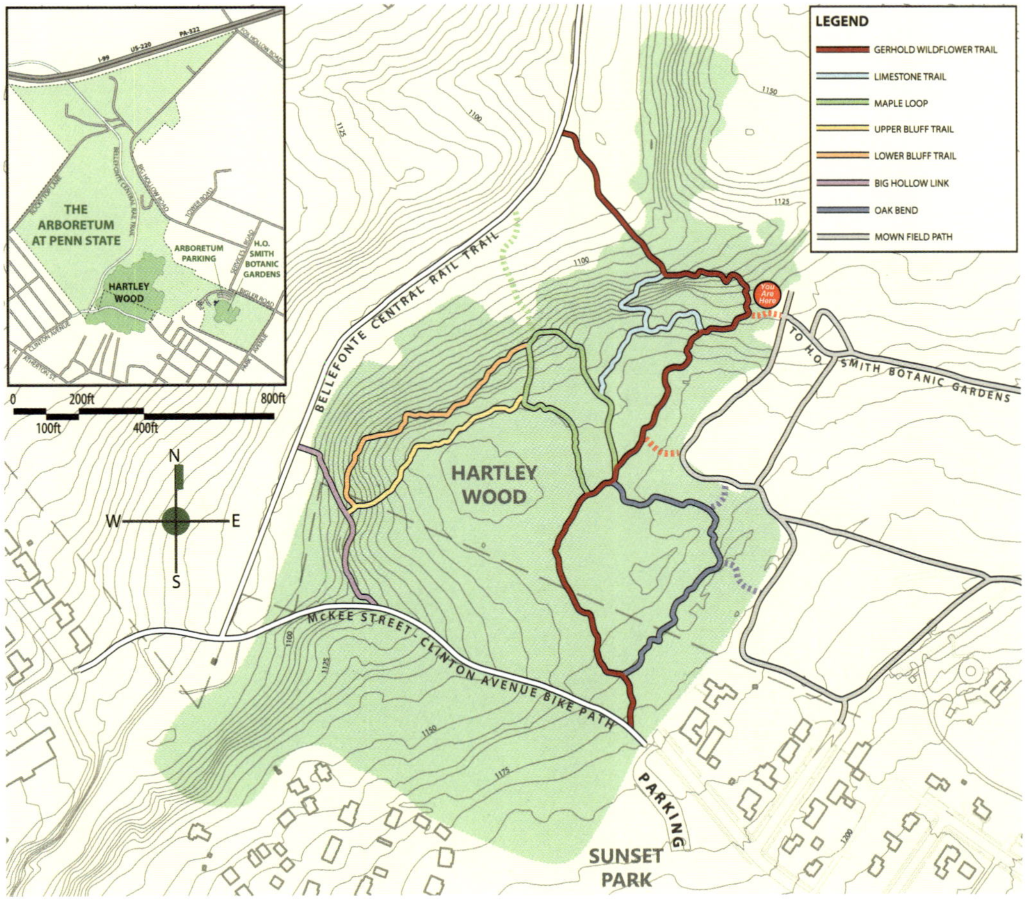

APPENDIX K

Rendering of the Gund Plan for an Education Center, Conservatory, and Planetarium (2012)

APPENDIX L

Plan for a Cultural District in the Arboretum by Ayers Saint Gross (2016)

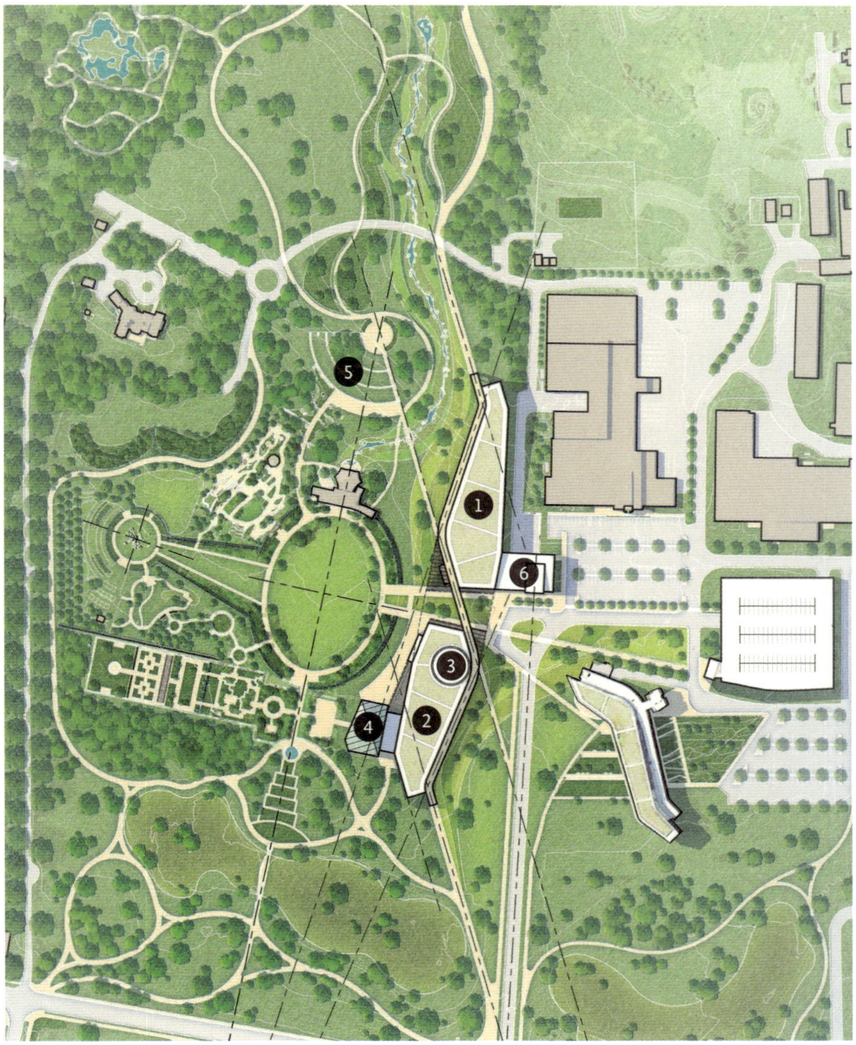

LEGEND

1 Art museum
2 STEM museums
3 Planetarium
4 Conservatory
5 Amphitheater
6 Classrooms & collaborative galleries

APPENDIX M

Schematic Design for a Children's Garden by Didier and AECOM (2010)

Overview of the Children's Garden design by Didier and AECOM.

Designer's rendering of the Fossil Ridge portion of the future Children's Garden.

Designer's rendering of the Mushroom Hollow, Discovery Tree, and Glass House.

Designer's rendering of the Children's Garden entrance and Transformation Canopy.

Designer's rendering of the Grotto (later redesigned as the Limestone Cave).

APPENDIX N

Plan for a Pollinator Garden by Oehme, van Sweden (2013)

APPENDIX O

Plan for a Pollinator Garden by M·T·R (2014)

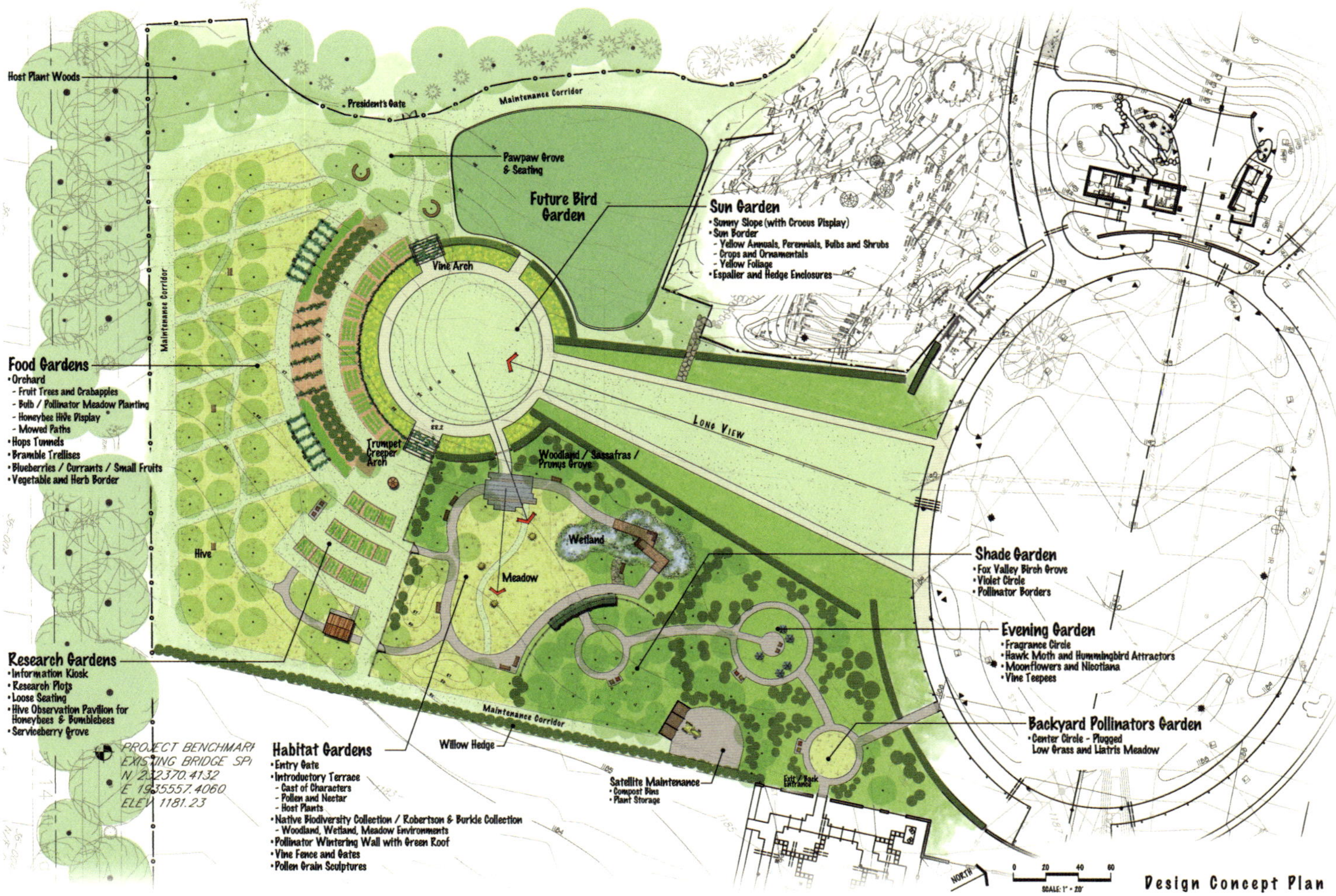

APPENDIX P

Final Pollinator and Bird Garden Design by Didier (2019)

LEGEND

1. Meadow Disc
2. Meadow Alcove
3. Shade Garden
4. Bird House and Bird Bath Water Feature
5. Bird Habitat Garden
6. Bird Pond and Wetland
7. Hilltop Overlook
8. Owl Reading Circle
9. Pond with Boardwalk & Stone Fence
10. Orchard Shade Trellis
11. Fruit Orchard
12. Pollinator Hotels
13. Event Patio
14. Evening Garden
15. Research and Demonstration Gardens
16. Honeybee Observation Hive

Soil zone plan for the Pollinator and Bird Garden.

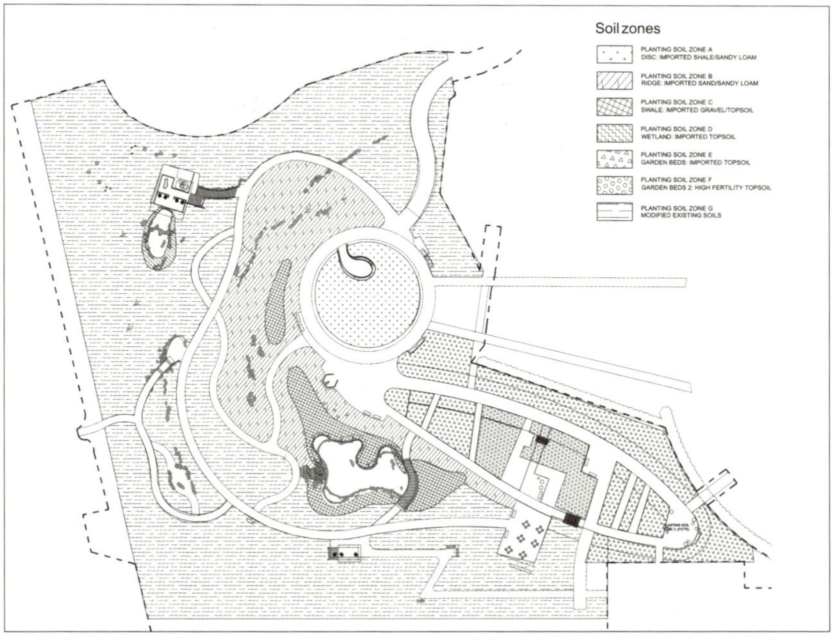

A sample of pollinator "stories" suggested by Claudia West of Phyto Studio based upon information provided by the project oversight team. These themes informed the selection and arrangement of plants.

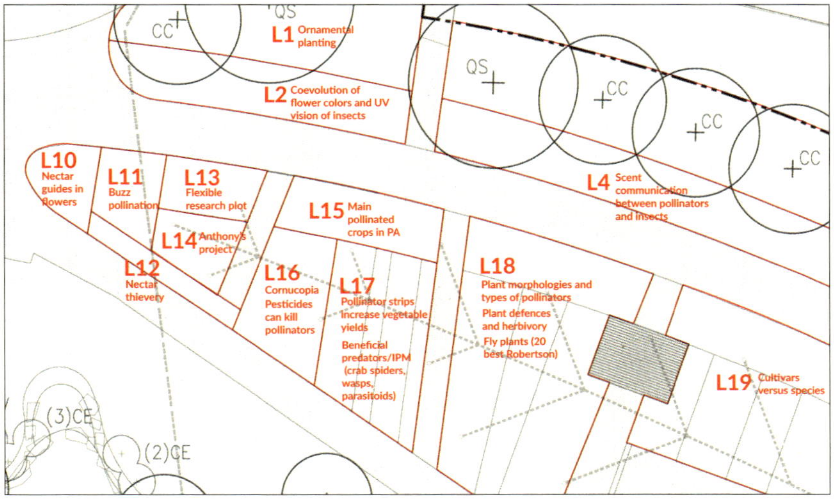

APPENDIX Q

Maps of the H. O. Smith Botanic Gardens and The Arboretum at Penn State

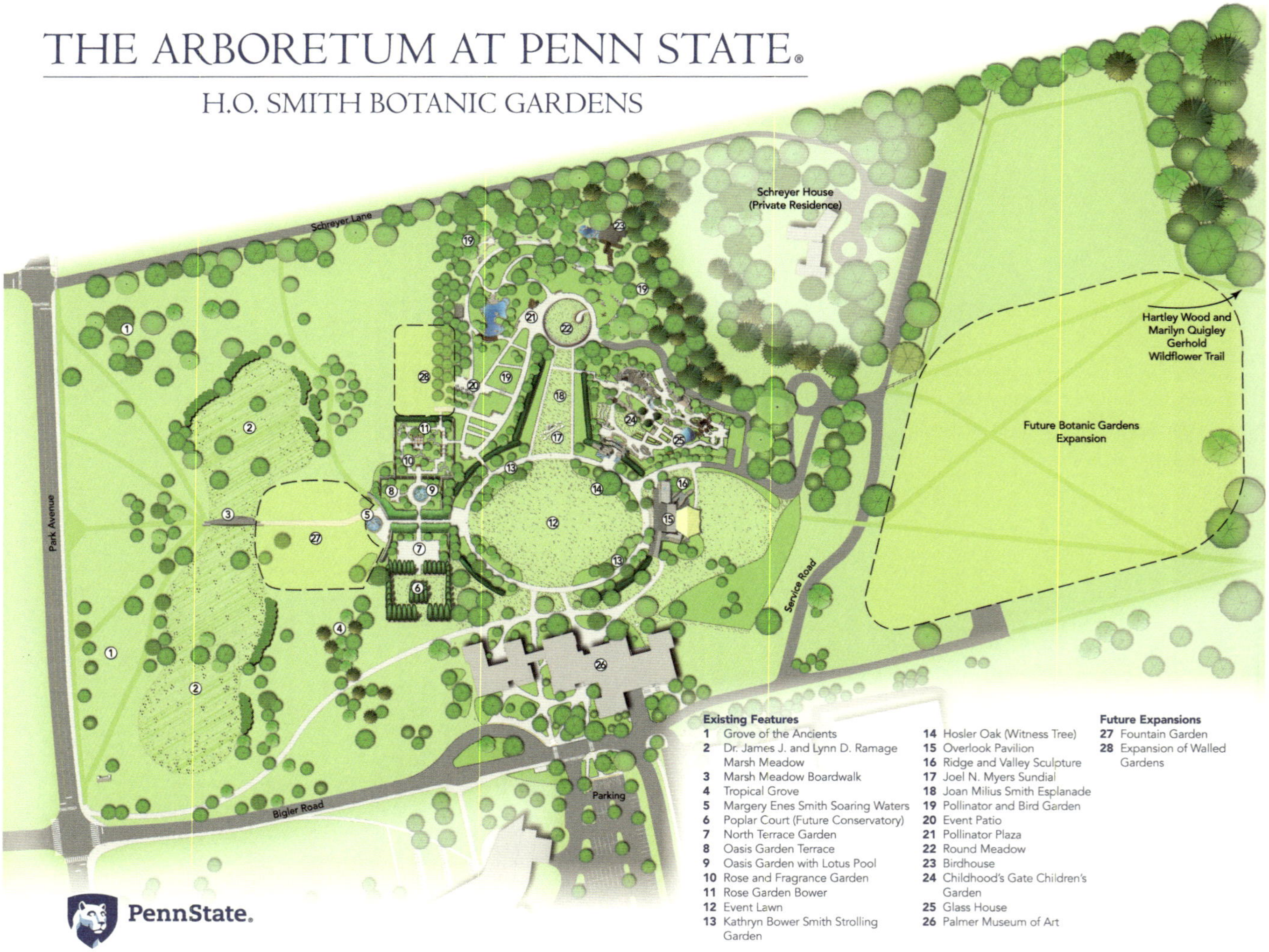

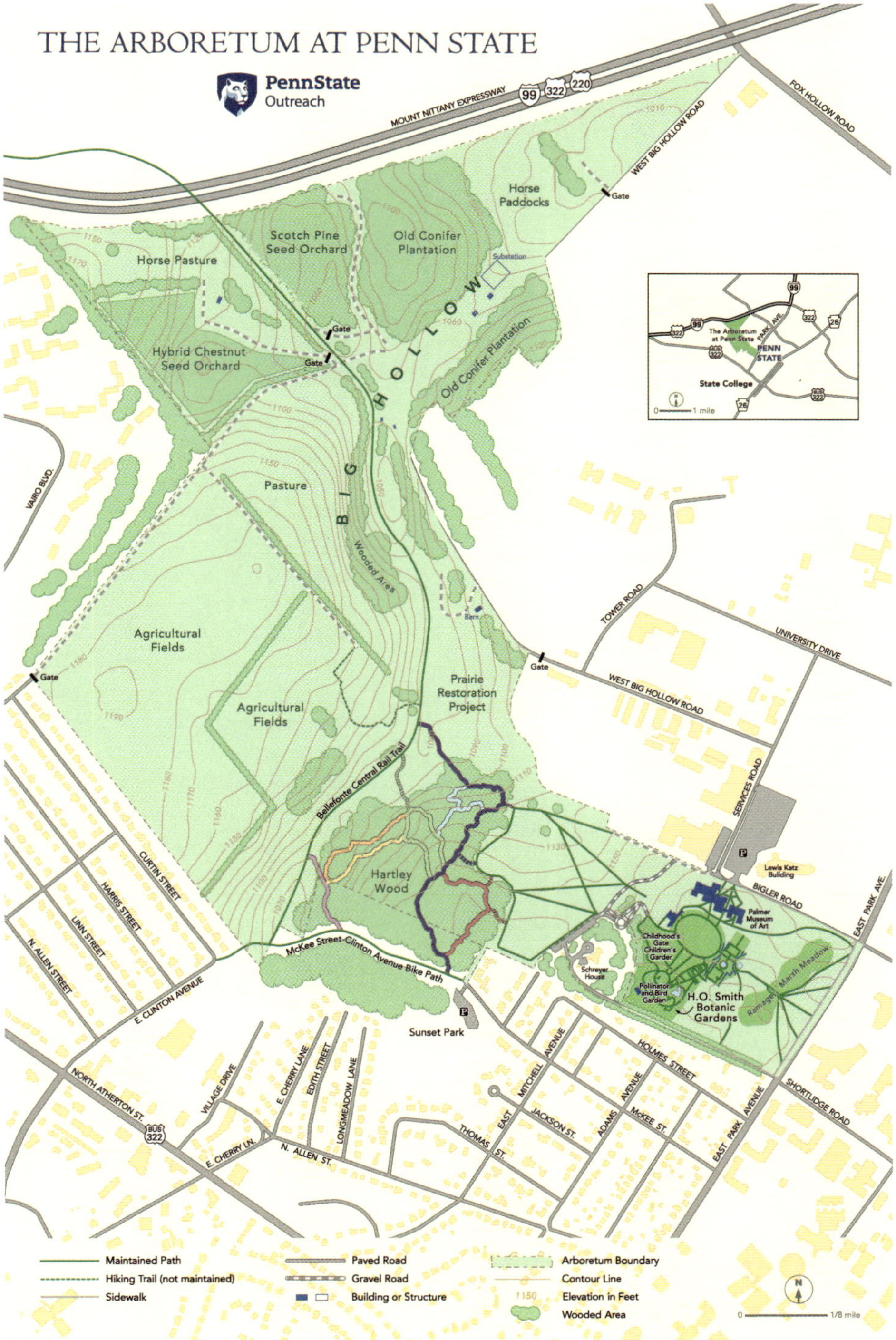
THE ARBORETUM AT PENN STATE
PennState
Outreach
MOUNT NITTANY EXPRESSWAY
99
322
220
FOX HOLLOW ROAD
WEST BIG HOLLOW ROAD
Gate
Horse Paddocks
Scotch Pine Seed Orchard
Old Conifer Plantation
Horse Pasture
Substation
BIG HOLLOW
Hybrid Chestnut Seed Orchard
Old Conifer Plantation
Pasture
VAIRO BLVD.
Wooded Area
Barn
Agricultural Fields
Agricultural Fields
Prairie Restoration Project
TOWER ROAD
UNIVERSITY DRIVE
WEST BIG HOLLOW ROAD
SERVICES ROAD
Bellefonte Central Rail Trail
Hartley Wood
Lewis Katz Building
BIGLER ROAD
Palmer Museum of Art
Childhood's Gate Children's Garden
Schreyer House
Pollinator and Bird Garden
H.O. Smith Botanic Gardens
Ramaley Marsh Meadow
EAST PARK AVE.
McKee Street-Clinton Avenue Bike Path
Sunset Park
CURTIN STREET
HARRIS STREET
LINN STREET
N. ALLEN STREET
E. CLINTON AVENUE
NORTH ATHERTON ST.
VILLAGE DRIVE
E. CHERRY LANE
EDITH STREET
LONGMEADOW LANE
E. CHERRY LN.
N. ALLEN ST.
THOMAS ST.
EAST MITCHELL AVENUE
JACKSON ST.
ADAMS AVENUE
McKEE ST.
HOLMES STREET
EAST PARK AVENUE
SHORTLIDGE ROAD
The Arboretum at Penn State
PARK AVE.
PENN STATE
State College
0 1 mile
Maintained Path
Hiking Trail (not maintained)
Sidewalk
Paved Road
Gravel Road
Building or Structure
Arboretum Boundary
Contour Line
1150 Elevation in Feet
Wooded Area
N
0 1/8 mile

119

Notes

Unless otherwise stated, individuals are identified by academic department or university unit.

A Long History of Proposals and Abandoned Efforts

1. Both "arboreta" and "botanic gardens" are curated plant collections, but they are not mutually exclusive, and one may contain the other. Nominally, arboreta are collections of trees only, but the term is often applied to large installations that also encompass botanic gardens, natural areas, conservatories, and other features.
2. As described below, Rausch was a business partner of Missy Marshall's; Missy became one of our planning consultants around the time of Rausch's retirement.
3. Arboretum Task Force members (1994–96) were Robert Berghage (Horticulture, replacing Perry Morgan in 1995), George Dickie (Landscape Architecture), John Joseph (OPP), Carl Keener (Biology), Perry Morgan (Horticulture), Robert Nuss (Horticulture), Eliza (Pennypacker) Oberholtzer (Landscape Architecture), Kim Steiner (School of Forest Resources), Kenneth Tamminga (Landscape Architecture), Stephen Wallner (Horticulture), and David Zehngut (OPP).
4. The memo was addressed to Peter Cavanaugh, chair of a University-level Life Sciences Study Group, and copied to Eva Pell, distinguished professor of plant pathology.
5. Early reports of the extent of this tract stated the acreage as 396. Later GIS work by one of my students, Samuel Grinstead, determined the area to be 370 acres based on original survey records. The tract outlined in the 1995 report is essentially that of the current arboretum (see appendix Q for a map).

The University Commissions a Formal Plan

1. Sasaki plan oversight team members were Robert Berghage (Horticulture), Richard Craig (Horticulture), Henry Gerhold (School of Forest Resources), John Joseph (OPP), Carl Keener (Biology), Ralph Mumma (Entomology), Robert Nuss (Horticulture), Eliza Pennypacker (Landscape Architecture), Robert Snetsinger (Entomology), Kim Steiner (School of Forest Resources), Andrew Stephenson (Biology), Glenn Steyers (Landscape Architecture), Kenneth Tamminga (Landscape Architecture), Christopher Uhl (Biology), and David Zehngut (OPP).
2. The Facilities Planning and Advisory Board was a committee of faculty and staff appointed by the assistant vice president for physical plant.
3. Three or four years later, an alumnus suggested we raise money by installing a columbarium (a place for funerary urns) in the arboretum, but that idea was quickly nixed by Old Main. At that time, at least, George Atherton was the only person whose remains were permitted on campus.
4. For reasons now forgotten, we began calling this a "preliminary master plan," and that is how the report is titled. However, except within the Mitchell Tract, the report has been the arboretum's only "master plan" for twenty-five years, and to avoid confusion that is what I have called it.

From Plan to Action, on a Wing and a Prayer

1. Arboretum Steering Committee members were Robert Steele (chair 1999–2009, dean, College of Agricultural Sciences), Bruce McPheron (chair 2009–12, dean, College of Agricultural Sciences), Barbara Christ (chair 2012–14, dean, College of Agricultural Sciences), Blannie Bowen (chair 2014–17, vice provost for

academic affairs), Kathleen Bieschke (chair 2017–21, vice provost for faculty affairs), Eric Barron (2002–6, dean, College of Earth and Mineral Sciences), Stephen Carpenter (2019–21, dean, College of Arts and Architecture), Douglas Cavener (2015–20, dean, Eberly College of Science), Richard Durst (2000–2006, dean, College of Arts and Architecture), John Dutton (1999–2002, dean, College of Earth and Mineral Sciences), William Easterling (2007–17, dean, College of Earth and Mineral Sciences), David Gray (2012–20, senior vice president for finance and business), Barbara "Bobbi" Korner (2007–19, dean, College of Arts and Architecture), Lee Kump (2018–21, dean, College of Earth and Mineral Sciences), Tracy Langkilde (2020–21, dean, Eberly College of Science), Daniel Larson (1999–2015, dean, Eberly College of Science), Kimberly Lawless (2020–21, dean, College of Education), David Monk (2019, dean, College of Education), Eva Pell (1999–2009, senior vice president for research), Neil Porterfield (1999–2000, dean, College of Arts and Architecture), Richard Roush (2014–21, dean, College of Agricultural Sciences), and David Wormley (1999–2013, dean, College of Engineering).
2. These included a checklist of plant species on the arboretum property (Daniel Laughlin, Biology), a cultural history of the property (Laura Enghofer, Landscape Architecture), a nature-restoration plan for a portion of Big Hollow (Laura Hamilton, Landscape Architecture), and a management plan for Hartley Wood (Samuel Grinstead, School of Forest Resources).
3. Faculty and Staff Advisory Committee members (aggregate for the years 2000–2010) were Robert Berghage (Horticulture), Daniel Carter (School of Theatre), C. Andrew Cole (Landscape Architecture), Robert Cooper (Energy and Engineering, OPP), Claude dePamphilis (Biology), Peter Everett (Marketing), Kathleen Fescemyer (Life Sciences Library), James Frazier (Entomology), Andrew Gapinski (arboretum), Alan Graefe (Recreation, Park, and Tourism Management), George Hamilton (Crop and Soil Sciences), Deryck Holdsworth (Geography), Carl Keener (Biology), Kathleen Kelley (Horticulture), Neil Korostoff (Landscape Architecture), Judith Larkin (Design and Construction, OPP), Robert Lima (Spanish, Italian, and Portuguese), Bruce Lindsay (Statistics), Jack Matson (Civil and Environmental Engineering), Eliza Pennypacker (Landscape Architecture), Kathleen "Kate" Reeder (arboretum), Joyce Robinson (Palmer Museum of Art), Peter Rubba (World Campus), Carolyn Sachs (Agricultural Economics and Rural Sociology), James Sellmer (Horticulture), Rudy Slingerland (Geosciences), Daniel Stearns (Horticulture), Andrew Stephenson (Biology), Elwin Stewart (Plant Pathology), David Sylvia (Crop and Soil Sciences), Kenneth Tamminga (Landscape Architecture), Alan Taylor (Geography), Donald Thompson (Food Science), Gordon Turow (Campus Planning and Design, OPP), Christopher Uhl (Biology), Paul Wangsness (College of Agricultural Sciences), Eileen Wheeler (Agricultural and Biological Engineering), Patrick Williams (arboretum), Scott Wing (Architecture), and David Zehngut (Campus Planning and Design, OPP).

Keeping the Faith Through Uncertain Times

1. Her graceful botanical drawings were later used for letterheads, business cards, brochures, promotional folders, and even T-shirt designs. They became an integral part of the arboretum's early graphic identity.

The Bellefonte Central Rail Trail

1. The Bellefonte Central Railroad provided regular passenger service to State College (with arrival at campus) from 1892 to 1946. Freight service to State College continued until 1974, when the line was abandoned.
2. Bellefonte Central Rail-Trail Project Management Team members (2000–2006) were Erik Brown (PennDOT), Glen Cauffman (College of Agricultural Sciences), Robert Cooper (Energy and Engineering, OPP), Douglas Erickson (Patton Township), John Gaudlip (Environmental Health and Safety, OPP), Rick Gilmore (Centre Rails-to-Trails Association), John Joseph (Campus Planning and Design, OPP), Judith Larkin (Design and Construction, OPP), Trish Meek (Centre Regional Planning Agency), Chris Price (College Township), Kathleen "Kate" Reeder (arboretum), Cassandra Schmick (Ferguson Township), Kim Steiner (arboretum), Randall Swope (College of Agricultural Sciences), Alex Tatanish (DCNR), Gordon Turow (Campus Planning and Design, OPP), Mark Whitfield (State College Borough), and Bruce Younkin (Penn State Transportation Services).

The Mitchell Tract Master Plan

1. The firm reorganized as M·T·R Landscape Architects, LLC, after the retirement of Geoff Rausch.

2. Committee members were Eliza Pennypacker (Landscape Architecture), Robert Berghage (Horticulture), Robert Cooper (Engineering Services, OPP), Peter Everett (Marketing), James Sellmer (Horticulture), Gordon Turow (Campus Planning and Design, OPP), Scott Wing (Architecture), and David Zehngut (Campus Planning and Design, OPP).
3. Arboretum Design Committee members (aggregate for the years 2005–21) were Erin Coe (Palmer Museum of Art), Gale Fulton (Landscape Architecture), Derek Kalp (Campus Planning and Design, OPP), Greg Kufner (Planning, Design, and Properties, OPP), Eliza Pennypacker (Landscape Architecture), Daniel Stearns (Horticulture), Gordon Turow (Campus Planning and Design, OPP), Steven Watson (Campus Planning and Design, OPP), Scott Wing (Architecture), and David Zehngut (Campus Planning and Design, OPP).

The Jump-Start Plan

1. According to Robert Steele, this sight line from Curtin Road was also preserved by shifting the Erickson Food Science Building (2006) farther east than was proposed by its architect. The Food Science Building is in the same block of buildings as Business and Forest Resources but farther south, adjacent to Curtin Road.
2. Morris Arboretum, Botanic Garden of Western Pennsylvania (now Pittsburgh Botanic Garden), and Lake Erie Arboretum.
3. Attendees were Kathleen "Kate" Reeder and Kim Steiner (arboretum), Robert Steele and Paul Wangsness (College of Agricultural Sciences), Colleen Lieberman (Dickinson School of Law), Richard Marini (Horticulture), and the following OPP staff members: Robert Cooper, Thomas Flynn, John Gaudlip, Derek Kalp, Judith Larkin, Jeff Spackman, Dick Tennant, Gordon Turow, and David Zehngut. I do not recall why I did not invite the Arboretum Design Committee *as a body* to this charette (some members were present), but it was probably because we were expecting only to identify the obvious design changes necessitated by the Katz Building design.
4. Penn State's flower trials project continues at the Southeast Agricultural Research and Extension Center in Lancaster County. A medieval garden attached to the gardens on the Katz site was moved to a plot of land behind the Housing and Food Services Building and has since been abandoned.
5. This represented a savings of about $12 million (in 2006 dollars) compared to the 2002 plan. One of our goals with the revision was to look for ways to reduce costs with minimal sacrifice in quality.

Our Witness Tree

1. Hosler served the University as a faculty member, department head, dean, senior vice president, and acting provost, and he was a well-known campus personage during my early years at the University. He passed away at age ninety-nine in 2023. George Biemesderfer had died three months earlier at age ninety-one.
2. Derek left the University shortly afterward to take another position but returned a few years later to assume his former position. During his time on the staff from 2001 to 2021 and beyond, Derek lent a hand enthusiastically to all arboretum projects and was especially instrumental in the design and construction of the Children's Garden and Pollinator and Bird Garden.

The End . . . and Then a Beginning

1. In email exchanges after the meeting among Rodney Kirsch, Graham Spanier, Gary Schultz, Rodney Erickson, Robert Steele, and me, it was agreed that the arboretum would be included as an Ag Sciences fundraising goal in the next capital campaign with a placeholder of $10 million. The "For the Future" capital campaign began in 2007 and ended in 2014. The $10 million would have been 10 percent of the College of Agriculture's fundraising goal for that campaign. If it succeeded, the College would probably not raise the $10 million in a single gift. Therefore, Graham stipulated that the project launch would be at the president's discretion and dependent upon a "cornerstone gift," and this uncertainty would be made clear to donors.
2. The provost's support was resumed a few months later when it became apparent that a gift was imminent.
3. When Jeff was still on the arborist crew, I happened to see him as he was taking down a large, dying tree between the Pavilion and Ferguson Building, where I had my office at the time. He was about twenty feet up in a bucket truck, lopping off branch sections with a small chainsaw. As each fell, he took one hand off the saw to nudge the section into a flat landing, avoiding small divots in the lawn. It was a minor incident, but I took note because it was unusual to see someone in that job taking, unobserved, so much *almost* needless care on his own initiative.
4. This became true, I believe. One obvious benefit was that our maintenance contract helped Jeff make the case in 2011 for a $2.9 million nursery and greenhouse plant production facility to enhance campus beautification efforts. Both the campus and the arboretum benefited from the cost savings and better

seasonal displays. The greenhouse also enabled us to overwinter large tropical plants for summer use in the arboretum.

5. Runyon projected an "opening year" revenue from admission fees of $287,000 (in 2002 dollars). We could not charge for admission without a way to control access to the gardens, which we did not have. Fortunately, with the operating endowment, the arboretum has been able to balance the budget without a gate fee, and I am glad. As explained below, the University eventually made a commitment to never charge for admission.
6. Capitalized names are key features in Missy's 2006 revision of the master plan (see appendix G).
7. The comedian Michael Feldman, doing a show in Eisenhower Auditorium on June 9, 2007, joked about Penn State "finally putting two trees together to form an arboretum." This alluded to the public announcement of Skip's gift, which would provide company for the "lonely" witness tree. I did not much care for the joke and neither did the president, as I learned many years later from someone who had been sitting near him. But it was clever and prompted loud laughter and applause from the audience.

The H. O. Smith Botanic Gardens

1. As described earlier, we had created a complete phasing scheme for the Mitchell Tract master plan, and it seemed natural to call this Phase I. However, new construction after Phase I followed according to our success in raising the funds for individual features, and we stopped speaking of phases.
2. It was not until after Phase I that we had the time and staff resources to develop a written collections policy. In 2009, I formed a Collections Committee, and the committee developed the first draft of a policy in 2011, which we continued to refine for several years. The committee provided guidance on questions of plant material acquisitions and donations, and it participated in revisions to the arboretum's mission and vision statements and other important matters. Members of the Collections Committee who served during the years 2009–21 were Alisha Barton (arboretum), Robert Berghage (Horticulture), C. Andrew Cole (Landscape Architecture), Claude dePamphilis (Biology), Amanda DiBella (arboretum), Shari Edelson (arboretum), Andrew Gapinski (arboretum), Rick Grazzini (Garden Genetics, Inc.), Richard Johnson (arboretum), Derek Kalp (OPP), Margot Kaye (Ecosystem Science and Management), Carl Keener (Biology), Harland Patch (Entomology), James Sellmer (Horticulture), Kim Steiner (arboretum), Andrew Stephenson (Biology), Alfred Traverse (Geosciences and Biology), and Peter Wilf (Geosciences).
3. P. S. Wyse Jackson and L. A. Sutherland, *International Agenda for Botanic Gardens in Conservation* (Richmond, Surrey: Botanic Gardens Conservation International, 2000); M. A. Jaramillo, P. Jimenez, and K. Frediani, "The Educational Garden: Exploring the Multifaceted Roles of University Botanic Gardens Through History," *Biological Diversity* 2, nos. 2–3 (2025): 85–94, https://doi.org/10.1002/bod2.70006.
4. The Grove of the Ancients encircles most of the Marsh Meadow on the Park Avenue lawn. It contains trees of genera that existed in the Eocene or earlier, roughly forty to two hundred million years ago, and would have been recognizable to a time-traveler.
5. The Tropical Grove is situated between the Fountain and Bigler Road. It contains regionally native tree species that are aberrant, northern outliers of tropical plant families, surviving relics from a time tens of millions of years ago when the region was much warmer. Not surprisingly, these trees tend to have a somewhat exotic look when compared with native trees descended from temperate-climate ancestors.
6. The occasion was marked with brief presentations by Robert Steele (as chair of the Arboretum Steering Committee), President Graham Spanier, James Broadhurst (chair of the Board of Trustees), and Kim Steiner.
7. It was called the Kathryn Bower Smith Strolling Garden after the donor's mother.
8. As our construction quality representative in OPP, Katie played a key role in ensuring contract compliance on the Phase I and Children's Garden construction projects. She later became the facilities manager for the College of Arts and Architecture and was on the planning committee for the Palmer Museum of Art.
9. The event featured presentations by President Graham Spanier, Steve Garban (chair of the Board of Trustees), Kim Steiner, Melissa "Missy" Marshall (as landscape architect), and Barbara "Bobbi" Korner (dean, College of Arts and Architecture).

The Arboretum Becomes Operational

1. Eventually, all operating funds for the arboretum came from endowment income (the largest source of funds), earned income from event rentals, annual giving (undesignated gifts), supplemental funds from the administration, and gifts designated for special programs, such as children's education and internships. Regular supplemental funding from the provost for salaries continued until 2014, when the administration began a

series of contributions to the arboretum endowment to offset and eventually cancel the need for an annual supplement.

2. The arboretum's mission statement was gradually refined in consultation with faculty and staff advisory groups beginning with a statement in the Sasaki master plan developed by that oversight team. The following statements were approved by the Arboretum Steering Committee on June 3, 2014.

   Mission: To engage the academic strengths of the University in promoting scholarship and education about plants and their history and importance on earth.
   Vision: The Arboretum strives to be a place of beauty and renewal, a venue for the arts, and a pathway to discovery and enrichment for all who visit.
   Core Values: We are dedicated to excellence, emphasizing the quality and richness of collections, materials, function, and design.

3. Accession records, including GPS coordinates, were completed in spring 2018 for every plant (or plant massing) in the H. O. Smith Botanic Gardens.

Incremental Enhancements

1. Dr. James J. and Lynn D. Ramage Marsh Meadow, after the donors. The Ramages made several generous gifts to the arboretum.
2. The arboretum was also selected as the beneficiary of the Class of 2019 gift, in this case to enhance one of the entrances to the Pollinator and Bird Garden.
3. The analemma is a lopsided "figure eight" described by the tip of a gnomon's shadow at a given hour, say noon, over the course of the year. This annual variance, east to west and back as the shadow lengthens and shortens, is caused by the Earth's tilt and elliptical orbit. Without accounting for the analemma—and most commercial sundials do not—the indicated time may be off by as much as a quarter hour.
4. Mark's work on the sundial helped get him selected to design the Class of 2011 gift, Veterans Plaza, next to Old Main. Mark, incidentally, knew the architect of the Katz Building, Richard Olcott, and delighted in sending him a photograph carefully framed to depict the gnomon of his sundial looming over Olcott's building (which at the time was still visible from the sundial) with the note, "Mine's bigger than yours."
5. Speakers were President Graham Spanier, Steve Garban (chair of the Board of Trustees), Kim Steiner, and Joel Myers (as the benefactor).
6. Another 13 acres of Hartley Wood is the forested part of Sunset Park, owned by the Borough of State College. Both parcels were part of the 467-acre warrant granted to James Hartley in 1784 (see appendix B).
7. Samuel C. Grinstead, "A Restoration Management Plan for the Hartley Wood in The Arboretum at Penn State" (M.F.R. paper, School of Forest Resources, The Pennsylvania State University, 2007).
8. Grinstead, "Restoration Management Plan."
9. See D. C. Laughlin, "Limestone Prairies and the Threatened Grass, *Bouteloua* curtipendula, in Pennsylvania" (M.S. thesis, Pennsylvania State University, 2002); this was published, in part, in D. C. Laughlin and C. H. Uhl, "The Xeric Limestone Prairies of Pennsylvania," *Castanea* 68 (2003): 300–316. For his assistantship responsibilities, Daniel conducted an inventory of the plant species that occur within the 370-acre arboretum footprint. His report, *Floristic Inventory of The Arboretum at Penn State*, Special Publication 1 (University Park: The Arboretum at Penn State, 2001), documents the occurrence of 410 plant species, of which 37 percent are not native to Pennsylvania.
10. We do not know whether a prairie community ever existed on the arboretum property, so our prairie work is better called a re-creation than a restoration.
11. The site is part of the "Foster Farm" (see appendix B) sold off by Centre Furnace owners in 1874 and acquired by the University in 1935. The foundation of the Foster farmhouse is within the "prairie" area and was partially excavated in spring 2015 and 2016 by Claire Milner and her students in ANTH 492 and ANTH 493.

A Home for the Staff

1. The oversight team members for what became the Gund plan (2011–13) were Robert Berghage (Horticulture), Claude dePamphilis (Biology), Jeffrey Dice (Buildings and Grounds, OPP), Andrew Gapinski (arboretum), Judith Larkin (Design and Construction, OPP), Michael Owen (Garden Genetics, Inc.), Christopher Palma (Astronomy and Astrophysics), Eliza Pennypacker (Landscape Architecture), Lawrence Ramsey (Astronomy and Astrophysics), Daniel Stearns (Horticulture), Kim Steiner (arboretum), Gordon Turow (Campus Planning and Design, OPP), Scott Wing (Architecture), and David Zehngut (Campus Planning and Design, OPP).
2. In 2015, we worked with a team of students in the College of Engineering's Learning Factory capstone course to develop and

install electronic visitor counters at three access points to the H. O. Smith Botanic Gardens. The counts were imperfect for various logistical reasons (people entered at other points, undercounts occurred when visitors were clustered as they passed sensors, etc.), but the estimate of 150,000 is based in part on those data and in part on data provided by the Centre County Convention and Visitors Bureau showing annual visitation by persons with active cell phones.

3. I had considered a similar plan as early as 2002, when I suggested to Dean Snow, department head in anthropology, that the anthropology and other campus museums could be relocated to a single facility in or near the arboretum. John Dutton, dean of Earth and Mineral Sciences, had earlier suggested something similar in connection with the EMS museum. At the time, I was interested in leveraging interest and investment in the arboretum, which of course was still unfunded in 2002. Dean Steele was leery of this approach, and we did not pursue it.
4. All arboretum planning, beginning in 2006, acknowledged that the land near Katz, Bigler Road, and Housing and Food Services was better suited for buildings and parking than gardens. However, in the context of the cultural district, it was my opinion that the land around any new buildings should be regarded and managed as part of the arboretum—hence my opposition to "subdivision" of the arboretum tract. As described in the next paragraph, President Barron's proposal that the cultural district be "a part of the arboretum" ratified this view of the project. Indeed, it could hardly have been intended otherwise: the arboretum was very popular, and the cultural district by contrast was expected to face heavy opposition from the Board of Trustees in part because of its cost.
5. Cultural District Plan Steering Committee members (2015–16) were Robert DeMayo (Transportation Services), Russell Graham (Earth and Mineral Sciences Museum), Deborah Howard (Facilities Resources and Planning, OPP), Barbara "Bobbi" Korner (College of Arts and Architecture), Rachel Prinkey (Design and Construction, OPP), Kim Steiner (arboretum), Gordon Turow (Campus Planning and Design, OPP), and Steven Watson (Campus Planning and Design, OPP).
6. The museum consortium encompassed a very disparate set of collections. Aside from conventional STEM museums there were, for example, the All-Sports Museum, the Pasto Agricultural Museum, and the Mascaro-Steiniger Turfgrass Museum of antique equipment. As it was, I believe these three preferred not to move to the arboretum, but they emphasize the thematic breadth of choices the planners faced.
7. Art Museum Project Charge Committee members (2018–21) were Erin Coe (Palmer Museum of Art), Shari Edelson (arboretum), Pamela Garbini (Facilities Resources and Planning, OPP), Harry Gebhardt (Design and Construction, OPP), Greg Kufner (Planning, Design, and Properties, OPP), Rachel Prinkey (Design and Construction, OPP), Monica Reed (Design and Construction, OPP), Katie Rountree (College of Arts and Architecture), Kim Steiner (arboretum), Neil Sullivan (Planning, Design, and Properties, OPP), and Steven Watson (Planning, Design, and Properties, OPP).
8. I am using here the planner's convention of "Penn State north," that is, the top of University Park maps, which is more accurately northwest.
9. As described earlier, the arboretum's annual attendance was about five times that of the former Palmer Museum. Even if attendance doubles at the new museum, as expected, it will still be a subordinate destination for visitors parking in the Katz lot.

The Childhood's Gate Children's Garden

1. Children's Garden project oversight team members (2008–14) were Linda Duerr (Child Development Laboratory, College of Health and Human Development), Andrew Gapinski (arboretum), Richard Johnson (arboretum), Derek Kalp (Campus Planning and Design, OPP), Judith Larkin (Design and Construction, OPP), Eliza Pennypacker (Landscape Architecture), Joyce Robinson (Palmer Museum of Art), Kathleen "Kate" Reeder (arboretum), Kim Steiner (arboretum), Gordon Turow (Campus Planning and Design, OPP), Wendy Whitesell (Bennett Family Child Care Center), Patrick Williams (arboretum), Scott Wing (Architecture), and David Zehngut (Campus Planning and Design, OPP).
2. When we finished the Children's Garden, Linda was directing educational programs in the Child Care Center at Hort Woods as an instructor in Health and Human Development.
3. Drawing upon direct and indirect knowledge of campus and community expertise and interest, we identified nearly thirty persons as "stakeholders" whose input during design development might be valuable. Most were able to participate in one or both workshops, and their comments were indeed very helpful.
4. Numerous donors contributed to the Children's Garden in addition to the Hintzes and Charles "Skip" Smith, raising the total gift level to $5.1 million.
5. The lead benefactors did not wish to have their names included within the garden's formal name, which by mutual agreement

became Childhood's Gate Children's Garden, a name suggested by one of the Hintzes' daughters, Ginny Hintz Remy.

6. Among these I must particularly acknowledge the generous assistance of David (Duff) Gold, Peter Wilf, and Mark Patzkowsky, faculty members in the Department of Geosciences, who gave generously of their time to guide us on matters pertaining to the geology and geologic history of central Pennsylvania.
7. Speakers were President Rodney Erickson, Keith Masser (chairman of the Board of Trustees), Kim Steiner, and Edward and Helen Hintz (as lead benefactors).
8. The occasion was marked with brief presentations by President Eric Barron, Keith Masser (chairman of the Board of Trustees), Kim Steiner, Edward Hintz, and three granddaughters of Edward and Helen Hintz. It should be noted that on all occasions where the other lead benefactor, Charles "Skip" Smith, might have made public remarks at a groundbreaking or dedication, he declined the invitation.

The Pollinator and Bird Garden

1. The oversight team members (2012–13) were C. Andrew Cole (Landscape Architecture), Claude dePamphilis (Biology), Jeffrey Dice (Buildings and Grounds, OPP), Andrew Gapinski (arboretum), Christina Grozinger (Entomology), Derek Kalp (Campus Planning and Design, OPP), Margot Kaye (Ecosystem Science and Management), Judith Larkin (Design and Construction, OPP), Harland Patch (Entomology), Kathleen "Kate" Reeder (arboretum), Kim Steiner (arboretum), Andrew Stephenson (Biology), Alfred Traverse (Geosciences and Biology), Gordon Turow (Campus Planning and Design, OPP), and Patrick Williams (arboretum).
2. See, for example, Douglas W. Tallamy, *Bringing Nature Home: How You Can Sustain Wildlife with Native Plants*, updated and exp. ed. (Portland, OR: Timber Press, 2009).
3. Other additions to this contract were a preliminary design (completed in 2017) for a garden that would extend from the Fountain to the Marsh Meadow and a revised planting plan (completed in 2015) for the Grove of the Ancients at the front of the arboretum, near Park Avenue. For the latter, Peter Wilf, a paleobotanist in the Department of Geosciences, provided indispensable help in identifying living tree genera that existed in the Eocene and earlier. The Grove of the Ancients is gradually being implemented with arboretum operating funds. The Fountain Garden is still a fundraising goal.

   We assembled two faculty and staff design oversight teams for M·T·R's pollinator and bird garden projects, but their memberships overlapped greatly. Together, their members were Blannie Bowen (Office of the Provost), Margaret Brittingham (Ecosystem Science and Management), Claude dePamphilis (Biology), Jeffrey Dice (Buildings and Grounds, OPP), Shari Edelson (arboretum), Derek Kalp (Campus Planning and Design, OPP), Judith Larkin (Design and Construction, OPP), Harland Patch (Entomology), Eliza Pennypacker (Landscape Architecture), Kathleen "Kate" Reeder (arboretum), James Sellmer (Horticulture), Daniel Stearns (Horticulture), Kim Steiner (arboretum), Andrew Stephenson (Biology), Gordon Turow (Campus Planning and Design, OPP), Patrick Williams (arboretum), Scott Wing (Architecture), and David Zehngut (Campus Planning and Design, OPP).
4. Pollinator and Bird Garden project oversight team members (2018–21) were Jason Beale (Shaver's Creek Environmental Center), Robert Berghage (Horticulture), Margaret Brittingham (Ecosystem Science and Management), Shari Edelson (arboretum), Tomas Carlo Joglar (Biology), Derek Kalp (Planning, Design, and Properties, OPP), Nick Kerlin (retired, licensed bird bander and amateur ornithologist), Greg Kufner (Planning, Design, and Properties, OPP), Judith Larkin (Design and Construction, OPP), Margarita Lopez-Uribe (Entomology), Harland Patch (Entomology), Eva Pell (vice president and dean emerita), Eliza Pennypacker (Landscape Architecture), Brian Phiel (Buildings and Grounds, OPP), Kim Steiner (arboretum), Andrew Stephenson (Biology), Neil Sullivan (Planning, Design, and Properties, OPP), Steven Watson (Planning, Design, and Properties, OPP), Patrick Williams (arboretum), and Scott Wing (Architecture).
5. There is still much to be learned about native pollinators. A Penn State study using statewide volunteers in 2021 and 2022 tallied 235 species of bees, 7 of which had not been previously known to occur in Pennsylvania. See N. E. Turley et al., "Bee Monitoring by Community Scientists: Comparing a Collections-Based Program with iNaturalist," *Annals of the Entomological Society of America* 117, no. 4 (2024): 220–33.

Transitions

1. Charles "Skip" Smith endowed the arboretum directorship with a $3 million gift just three months before my retirement in 2021. Skip passed away in 2024 at age ninety-eight and left a bequest

to the arboretum in the neighborhood of $25 million. He was our great benefactor, and without his generosity I do not believe the arboretum would ever have been built.

2. Outreach is also the administrative home for Shaver's Creek Environmental Center. In November 2018, most of our staff made a field trip to Shaver's Creek to spend the day exchanging information about ourselves, our programs, and our budgets and administration. We had done a few joint public events with them previously, and the objective of this meeting was to look for opportunities for further collaboration in programs for the public. Their director, Mark McLaughlin, and I agreed to get our people together again. Unfortunately, the distraction of the Pollinator and Bird Garden construction beginning in 2019, COVID-19 in 2020, and the administrative transitions in 2021 came before our meeting could bear fruit.

Appendix B

1. Except where indicated, the information in appendix B was extracted from Daniel Laughlin, *Floristic Inventory of The Arboretum at Penn State*, Special Publication 1 (University Park: The Arboretum at Penn State, 2001); Laura Enghofer, *The Pennsylvania State University Arboretum Landscape History*, Special Publication 2 (University Park: The Arboretum at Penn State, 2002); and Samuel C. Grinstead, "A Restoration Management Plan for the Hartley Wood in The Arboretum at Penn State" (M.F.R. paper, School of Forest Resources, The Pennsylvania State University, 2007).

   The Enghofer report was updated in an unfinished 2012 manuscript by Kathleen "Kate" Reeder entitled "The Cultural Landscape of The Arboretum at Penn State: A History of Landowners and Land Use," which contains maps prepared by Grinstead and other new information. These reports are archived in the Penn State University Libraries.
2. Information about Native inhabitants of the area is from Randall M. Miller and William A. Pencak, eds., *Pennsylvania: A History of the Commonwealth* (University Park: Penn State University Press, 2002).
3. The arboretum lies just south of the northern boundary of the "reduced" tract, assuming that the boundary runs due west from its northeastern corner at 40° 49.435' N, 76° 50.813' W. Land to the immediate north of the Albany Purchase was acquired by the Pennsylvania colony through an additional purchase in 1768.
4. In fact, although Hartley had nominally purchased the property with the 1784 warrant, he held no legal title to it for fifty-three years, as he had failed to file for a patent on the property. The land was resurveyed in 1837 following the original traverse bearings, and the patent was issued to Krumrine upon the sale of the property. The resurvey indicated an area of 467 acres, more than had been calculated in 1784.
5. Roger L. Williams, "Building and Opening the Farmers' High School, 1855–59," *Mansion Notes* (Centre County Historical Society) 44, no. 3 (2022): 1, 12–13; Roger L. Williams, "Samuel Miles: Frontier Soldier, Revolutionary War Colonel, American Patriot," *Mansion Notes* (Centre County Historical Society) 47, no. 3 (2025): 3, 20–23.
6. There are several published histories of the Bellefonte Central Railroad and extensive archival materials about the railroad in the University Libraries. A useful summary with citations may be found at "Bellefonte Central Railroad," The Pennsy Modeler, https://jbritton.pennsyrr.com/index.php/post-pennsy/bellefonte-central-railroad.
7. See D. D. Stevenson, H. A. Meyer, and L. J. Bennett, *Management Plan for College Farm Woodlands* (n.p., 1939).
8. The acreage numbers here do not quite total because they are rounded.
9. Germination dates were extrapolated from increment cores taken by Grinstead near the base of the tree. The three seventeenth-century trees are in the Sunset Park portion of the woodlot.

Appendix F

1. These principles are based on concepts in the *Preliminary Master Plan for The Penn State Arboretum* by Sasaki Associates, Inc., and the *Master Plan for the Mitchell Tract in The Arboretum at Penn State* prepared by Marshall·Tyler·Rausch, LLC, as articulated by the Arboretum Design Principles and Guidelines Task Force.

Index

Numbers in **bold** refer to images in the gallery after page 50. *Italic* page numbers refer to images in the appendixes. Notes are indicated by "n" followed by the note number.